TAKE-OFF

Technical English
for Engineering

Course Book

David Morgan
and **Nicholas Regan**

3230211947

SLT

Garnet
EDUCATION

Published by

Garnet Publishing Ltd.
8 Southern Court
South Street
Reading RG1 4QS, UK

ISBN-13: 978 1 85964 974 9

British Library Cataloguing-in-Publication Data
A catalogue record for this book is available from the British Library.

Production

Project manager:	Richard Peacock
Project consultants:	Fiona McGarry, Rod Webb
Editorial team:	Carol Rueckert, Emily Clarke
Design:	Henry Design Associates, Robert Jones, Christin Helen Auth
Illustration:	Doug Nash
Photography:	Bob House; Clipart.com; Corbis; Getty Images; Stockbyte; www.boeingimages.com.
Audio production:	Motivation Sound Studios

Garnet Publishing wishes to thank Chris Murray and the staff of Saudi Development and Training (SDT) for their assistance in the development of this project.

Images on page 10 reproduced with kind permission of www.futureflight.org (University of Southampton).
Extract on page 92 reprinted with kind permission of *Shoreham Airport News*.

Every effort has been made to trace the copyright holders and we apologize in advance for any unintentional omissions. We will be happy to insert the appropriate acknowledgements in any subsequent editions.

Printed and bound

in Lebanon by International Press

Contents

Book map iv

Introduction vi

Unit 1 2

Unit 2 22

Unit 3 42

Unit 4 62

Unit 5 82

Unit 6 102

Unit 7 122

Unit 8 142

Unit 9 162

Unit 10 182

Unit 11 202

Unit 12 222

Additional material 242

Glossary 254

Electrical symbols 263

Word lists

Unit by unit 266

Alphabetical 276

Tapescript 284

Book map

Unit	Topics	Skills	Language
1 **Design and innovation**	• Properties of materials • Design rationale • Aircraft specifications	**Listening and reading:** • Develop sub-skills: skimming and scanning • Find key information from different sources **Speaking and writing:** • Descriptions and comparisons • Note completion	• Numbers and measurement • Language for description, e.g., material properties • Question forms • Reason and purpose • Tenses: past, present and future
2 **Manufacturing techniques**	• Functions of hand and machine tools • Manufacturing processes	**Listening and reading:** • Identify key information • Transfer information, e.g., from text to table or diagram **Speaking and writing:** • Explanation of components and processes • Contrast and comparison	• Definitions and descriptions • Verbs for manufacturing operations • Imperatives for instructions • Prepositions of movement • Word-building • Parts of speech
3 **Frameworks**	• Aircraft structure • Principles of flight • Assembly line jobs and processes	**Listening and reading:** • Practise extracting key information • Recognise 'clues' such as signposts in a text **Speaking and writing:** • Restate and summarise information • Practise note-taking and using notes for a talk	• Movement and manoeuvres 1 • Forces and stresses • Direction and location • Passive voice for processes • Speculation and prediction • Word combinations • Vowel sounds / sentence stress
4 **Control systems**	• Hydraulics • Hydraulic applications • Braking and landing systems • Aircraft control surfaces	**Listening and reading:** • Intensive comprehension: follow descriptions of procedures and systems **Speaking and writing:** • Describe, evaluate and compare systems • Practise reducing and summarising information	• Hydraulics, control surfaces and linkages • Movement and manoeuvres 2 • Adverbs of manner, degree and frequency • Compound nouns • Word stress and vowel strong / weak forms
5 **Engine and fuel systems**	• Engine parts and how engines work • Engine types and specifications • Engine overhaul • Fuels and fuel systems	**Listening and reading:** • Familiarisation with different types of text, e.g., data sheets and tables • Focus on text organisation **Speaking and writing:** • Discussion and speculation using notes, tables and diagrams • Practise note-taking using reduced forms and abbreviations	• Engine parts and operations • Verbs for engine problems and repair • Fuel and air movements • Mathematical concepts, symbols and abbreviations • Passive with *by* + agent • Reduced passive forms in notes
6 **Review unit**			

Unit	Topics	Skills	Language
7 Safety and emergency	• Risks and hazards • Emergency procedures and equipment	**Listening and reading:** • Familiarisation with longer and multiple texts • Work out meaning from context **Speaking and writing:** • Complete tables and reports with notes • Practise giving safety information and explanations	• Chemicals and elements • Safety warnings, equipment and systems • Nouns and adjectives for damage and dangers • Imperatives and modals of obligation • Language of purpose • Syllables and stress
8 Air and gas	• Pneumatics • Heating and cooling systems	**Listening and reading:** • Familiarisation with different text types, e.g., instruction manuals and advertisements • Interpret graphs and diagrams **Speaking and writing:** • Discuss and interpret diagrams and schematics • Form and table completion	• Measurements / calculations • Language for changes of physical state, e.g., *condense* • Technical verb / noun collocations • Compounds and complex noun phrases • Defining relative clauses • Discourse markers • Rhythm and stress
9 Electrical systems	• Electrical maintenance and repair • Electric motors and batteries • Diagrammatic representation	**Listening and reading:** • Interpret / follow explanations of charts and formulae • Recognition of words and sounds in a flow of connected speech **Speaking and writing:** • Explain and discuss electrical problems and procedures • Form-filling and detailed note-taking	• Electrical parts and systems • Electrical abbreviations and symbols • Multi-word verbs • Comparison and recommendation • Past and past perfect tenses • Consonants and clusters
10 Communication	• Avionics • Cockpit instruments and navigation systems • Technical drawing and PCBs	**Listening and reading:** • Practise skimming and scanning more complex texts • Follow instructions and detailed information **Speaking and writing:** • Develop editing skills	• Technical verb / noun collocations • Language for electrical faults and problems • Linking words • Fact, tendency and possibility: zero and 1st conditional sentences • Modal verbs for speculation
11 Maintenance	• Forms and certification • Maintenance procedures	**Listening and reading:** • Interpret and understand various forms • Ask questions and give explanations about maintenance systems **Speaking and writing:** • Discuss possible maintenance scenarios / problems • Practise writing clear instructions	• Language for position, assembly and disassembly • Language for mechanical damage • Word-building • Hypothetical situations: 2nd and 3rd conditional sentences • Phrases for explanations and definitions
12 Review unit			

Introduction

Who is *Take-off* for?

Take-off is aimed at people who need to study technical English. It was written primarily for engineering students whose English is intermediate level (European Framework B1) or above, but it is suitable for pre- and in-service technicians in all MRO areas (NVQ Level 2 and above). It covers general engineering topics, but has an aeronautics focus, so is also particularly suitable for anyone working in the aeronautics industry from co-MRO operatives to supervisors, managers and pilots.

The approach

Take-off assumes that you have a basic grounding in English grammar and some knowledge of technical terms, but that you need to improve your listening, speaking, reading and writing skills. Consequently, the book takes a *skills-building approach*, but will also help you develop the range of vocabulary and grasp of language structures relevant to your daily lives. In addition, it will help you develop 'survival' English so that you can pursue technical courses in further or higher education institutions, or start working as a technician or engineer in an English-speaking environment.

Features of the course:

- Language boxes: provide short, visual explanations of grammatical rules and syntactic patterns. You are encouraged to 'notice' the language patterns that are used in the texts, so new language is always presented in context.

- Skills boxes: focus awareness on the useful sub-skills and strategies that will help you in the real world. These include advice on different ways of reading, strategies for listening and advice on how to take notes and edit written work.

- Authentic texts: include instruction pages and diagrams from BA Systems manuals and certification and report forms. The use of this type of material will prepare you for the work environment and promote credibility.

- Learner-centred tasks: encourage you to draw on your own experience and contribute your own ideas in many activities. A lot of the tasks in *Take-off* involve pair or group work, including discussions, working out problems, presenting and swapping information. This maximises practice opportunities and facilitates a positive group dynamic.

- Phonology work: addresses specific difficulties which research indicates are critical. These include sentence stress, connected speech, vowel quality and length and consonant clusters.

- Word lists: include a list of key vocabulary for each lesson as well as an alphabetical list of all vocabulary at the back of the book. You will be able to check the words using technical dictionaries as preparation for the lesson.

- Glossary: a list of key technical terms is included at the back of the book. This is a useful resource which provides simple explanations and definitions.

- The Workbook: provides additional reading texts for comprehension practice as well as writing tasks and extra vocabulary and grammar-focused exercises. It may be used in class to provide extra written practice and language consolidation. Alternatively, your teachers may wish to concentrate on new vocabulary work and skills-building during class time, and set the Workbook activities for self-study.

- Independent learning: is encouraged through the use of tasks that require you to research topics in reference books and on the Internet. The Teacher's Book suggests web links for many of the lessons.

- The Teacher's Book: includes full, step-by-step procedures for each lesson as well as a full answer key and model answers for exercises where you are asked to write your own texts. It also includes advice for your teachers on how to clarify problem areas, suggestions for extra activities and links to useful websites.

The syllabus

Vocabulary

Vocabulary is taught systematically in clear contexts with a focus on collocation and word-building as well as spelling and pronunciation. High-frequency vocabulary has been selected from technical word lists, and semi-technical vocabulary is given particular prominence. Vocabulary tasks encourage you to group and categorise lexis according to topic and to focus on how vocabulary is used in a range of texts.

Vocabulary tasks include dictionary work, matching words with definitions, production of personalised sentences, and labelling diagrams and schematics. It is assumed that you will have your own technical dictionaries and/or access to class sets of dictionaries.

Grammar

Take-off focuses primarily on language that is used frequently in a technical environment. Therefore, it gives special emphasis to the following: imperatives, the passive voice (including reduced passive forms necessary for check lists) and modal verbs for obligation.

You are also encouraged to notice and learn 'chunks' of language and notional/functional exponents for key areas such as: talking about cause and effect, giving warnings, explanations and instructions, discussing how to deal with problems, calculations, measurements and numbers.

Listening

Three or four lessons in each unit are generally listening-focused. Recordings introduce learners to different types of English (including American and Australian varieties) and a range of regional accents.

You are encouraged to develop your ability to listen for gist and to pick out specific information. *Take-off* also helps you to develop micro-skills; these listening tasks help you to pick out keywords and signposts, recognise use of intonation, rhythm and pauses, and identify stressed words and syllables.

Speaking

Take-off aims to develop your confidence as well as your competence in speaking. Speaking activities are a central part of every lesson, and it is essential that you engage and participate in pair and group work. Tasks often require a degree of critical thinking in response to 'what do you think?' type questions. This will be vital to your further studies and training.

Tasks include: discussions, role-plays, information-gap exercises, giving short talks involving explanations and instructions, and solving problems. These are all useful preparation for the engineering workplace.

Speaking exercises are coupled with a focus on pronunciation: common L2 speaker problems such as word and sentence stress, vowel length and consonant clusters are given attention in more controlled speaking exercises.

Reading

Take-off helps you to develop skills required to deal effectively with all the text types you will come across in your training and workplace. These skills include: dealing with numbers, extracting data from dense technical text, working with bullet-pointed and numbered manual-style instructions, and reading and interpreting diagrams, flowcharts, graphs and labelled drawings.

Writing

Writing tasks are devised to meet your needs in both training and workplace situations. Tasks are scaffolded so that you can use the prompts and model texts to help you produce your own texts. Task types include: note-taking (including abbreviations), writing numbers and bulleted notes, form-filling, and transferring data from listening to pictorial, from diagram to notes and instructions, and from written text to speaking.

Bright ideas

Listening

1 Match the words in the two columns below, then label each picture with one of the phrases.

air ——————————— jet

flat screen ————— engine

bagless ——————— cushion

MP3 ————————— (vacuum) cleaner

vertical take off ——— television

high-performance ——— player

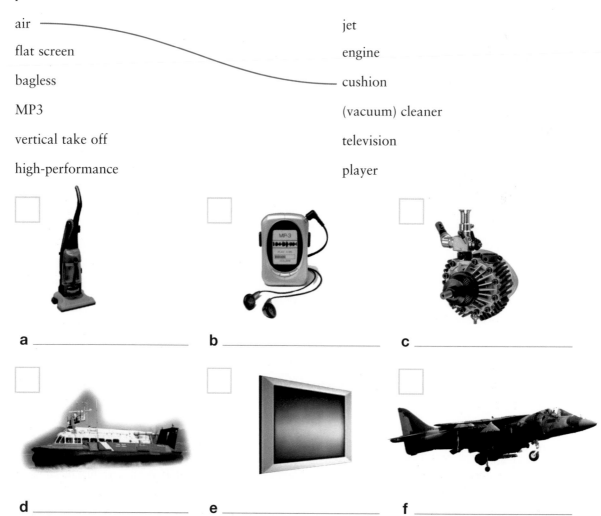

a _____ b _____ c _____

d _____ e _____ f _____

2 You are going to hear some short conversations about the technologies in Exercise 1. Listen and number the pictures according to the order in which they are discussed.

3 Listen again and circle the words below which you hear in the conversations.

picture	glass	memory	coffee	boat
water	music	shopping	base	driver
wings	machine	motor	motorway	

4 🎧 **Listen again. Fill in the table with detailed information.**

	product	measure	unit	how much?
a		weight		
b				1
c	MP3			
d			km/h	
e		capacity		
f				230

5 **Look at the completed table. Do you remember all the information about each product?**

Speaking

1 **With a partner, discuss the following questions. Use the phrases in the box below.**

   a Why do you think the products in the pictures are different from what came before?

   b Which technologies do you find interesting?

> *It's really* + (adj)   *It's got* + (noun)
> *It can* + (verb)   *You can* + (verb)

2 **Can you think of innovative technologies in the following areas? Discuss them in groups of three or four.**

- fuel sources
- IT and communications
- ways of identifying people
- health and medicine
- materials and textiles

3 **Each of you should choose a technological idea or invention that is innovative or important for the future.**

   a Imagine you have been asked to give reasons why your chosen technology would be the best one to spend a large Research & Development grant on. Make some notes.

   b When you are ready, talk with your group members. Try to persuade them that your ideas are the best.

   c When you have all spoken, decide who should win the grant.

➡ **Workbook pages 2/3**

Skills Box

Listening in different ways

To develop your listening skills, you need to be aware that we do not always listen in the same way.

In Listening Exercise 2, you listened to get the overall idea of the conversation. This involves using your knowledge of the topic and words that are connected with it. This is called *listening for gist*.

In Listening Exercise 3, you listened for single words, and in Listening Exercise 4 you identified more key information, e.g., numbers, dates. This is called *listening for specific information*.

Think about how you would listen to the following things in English:

- an airport announcement
- a lecture

Lesson 2
Product description

Reading

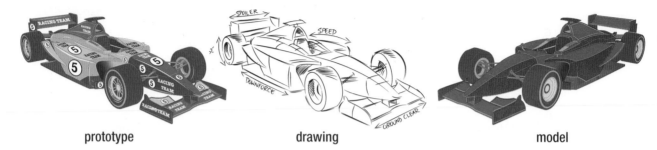

prototype   drawing   model

1 **What are the main stages of the design of a new product, from the first idea to the final finished product?**

2 **Look at the text below and make a list of the main stages.**

Everything that we make, from the largest airliner or military aircraft to the smallest **component** in a washing machine, starts life as an idea. However, there are many stages in the development process between the initial idea and the finished product. When deciding on the preliminary design, you may start with a **drawing**. More often,
5 though, the first stage involves doing some design research in order to come up with a design **specification** by asking a series of questions, such as:
How big is the product going to be?
How many are needed?
What **materials** will be needed to make it?
10 Who will use it?
What **conditions** will it have to withstand?
How much will it cost?

The answers to questions like these provide the designers and draughtsmen with a design brief. This gives them a clear plan to work with. Unfortunately, the design brief
15 sometimes changes in the middle of the project because, for example, the prices of materials increase, or because there are new safety **regulations**.

When the design brief is ready, the designers and draughtsmen have to come up with **sketches**, drawings and **models**. These show what the finished product will look like as well as providing the necessary information to make a **prototype**. The company
20 engineers then test the prototype very carefully to make sure that it follows the design brief. If there are problems, they can make changes before starting **mass-production**.

3 **Compare your list with a partner. Did you come up with the same stages?**

4 In the passage you just read, ten key words are shown in **bold**. Match each word to one of the definitions below.

a _____ (*n*) production of large quantities of something

b _____ (*npl*) small-scale versions of a product

c _____ (*npl*) situations and states, e.g., temperature

d _____ (*npl*) rough drawings or outlines

e _____ (*n*) a full-scale model or early version of the product

f _____ (*n*) a description of what the product will be like

g _____ (*npl*) rules

h _____ (*n*) detailed picture

i _____ (*n*) a part of a product

j _____ (*npl*) what a product is made of

Language

1 Look at Language Box 2. Circle all the pronouns *it, they, this* and *these* in the text in the Reading passage opposite.

2 What do these pronouns refer to in the text in the Reading passage?

a it (line 10) _____

b this (line 14) _____

c these (line 18) _____

d it (line 20) _____

e they (line 21) _____

3 Look at these examples of product descriptions from Lesson 1 and correct the mistakes in the use of pronouns.

a The Wankel rotary engine has neither a crankshaft nor a flywheel. These means that this is lighter and can produce more power than a normal engine of the same size.

b The ACV (hovercraft) flies on a cushion of air a short distance above the ground. It means that this can move easily on land or water.

⇨ Workbook pages 3/4

Lesson 3
Materials and properties

Vocabulary and speaking

In this lesson, you are going to think and talk about the materials that aircraft and other products are made of.

1 **Look around the classroom and try to identify all the different materials you can see.**

   a If you don't know the English name for all the materials, try to describe them with words you do know. This is called *paraphrasing* and is a very useful skill. For example: *It comes from a tree. It's used for making car tyres*. Look at the Language Box for some useful phrases.

   b Are any items made of the following: moulded plastic, copper, glass, metal alloy, paper, wood?

2 **Match the following questions and answers. What object is being described?**

Question	Answer
a What's it made of?	i It's about 10 cm by 10 cm.
b Can it be used for storing things?	ii Yes, it is.
c Is it hard?	iii Plated metal alloy.
d What is it for?	iv You turn it with your hand.
e How does it work?	v No, it can't.
f How big is it?	vi It's for opening a door.

3 **With a partner, discuss the questions below.**

   a Think of a car. How many different materials are used in its manufacture? Why is each material used?

   b What about an aeroplane? Look at these three pictures of aeroplanes and discuss the different materials you think you would find in them and why.

Reading

1 **Take two minutes to quickly read the following text. Decide which of the sentences below is the best summary of the whole text.**

    a The modern jet fighter is made using the most advanced materials.

    b The properties of materials are a central part of design.

    c Modern materials have completely replaced traditional ones.

Materials and design

As well as designing the structure, shape and size of a product, the designer must also specify the materials that it will be made of. In the past, the range of materials was very limited. Natural materials often had to be used because synthetic materials were not
5 available. Wood, steel, leather, cotton and glass were used to manufacture early cars, trains and aircraft as well as household products such as vacuum cleaners, radios and even TV sets! Of course, these traditional and natural materials are still used today, but the designer now has a wide range of synthetic materials to choose from as well.

A good example of the way in which new synthetic materials are used is the jet fighter in
10 the picture. The wings of this state-of-the-art fighter aircraft are made of carbon fibre composite and the nose cone is made of glass-reinforced plastic, which is also a composite material. This means that over 70 per cent of the aircraft body is non-metallic.

The choice of suitable materials depends on three main factors. These are suitability, availability and cost. In order to decide if a material is suitable, a designer has to know
15 about the physical and mechanical properties of a material. These are:

Physical properties	**Mechanical properties**
coefficient of linear expansion (how much you can stretch it)	toughness
	elasticity
specific heat capacity	strength
20 melting point	brittleness
conductivity	malleability
electrical resistivity	ductility
density	lustre
hardness	

25 The designer will also need to know about the chemical stability of the material under specific operating conditions. For example, will the material corrode if it is exposed to rain?

2 **Match the following questions with some of the mechanical properties listed in the reading passage. Then work in groups to think of more questions.**

    a Can it be used to make electrical wire?   _____

    b Will it break into pieces easily when you hit it suddenly?   *brittleness*_____

    c Can it be pulled into a long, thin shape?   _____

    d Does heat pass through it quickly?   _____

⇨**Workbook pages 4–6**

Lesson 4
An amazing material

Listening and note-taking

You are going to listen to an extract about spider silk from a popular science programme.

1 🎧 **Tick the topics from the following list which the speakers mention.**

a ☐ how spiders make silk

b ☐ the tensile strength of steel

c ☐ airliner design

d ☐ properties of spider silk

e ☐ elasticity

f ☐ difficulties for commercial use

g ☐ synthetic spider silk

h ☐ how spiders use their silk

i ☐ the clothes industry

2 Now listen for key information.

    a First, look at the Skills Box.

    b Copy the topics from Exercise 1 that the speakers mentioned into the space below. Leave space after each topic for writing notes.

    c 🎧 Listen to the programme again and make notes beside each topic.

3 Compare your notes with a partner. How much of the key information can you remember between you?

> ## Skills Box
>
> ### Taking notes
>
> You will need to take notes quickly at lectures and meetings. These notes must be clear enough for you to understand later.
>
> - Don't write full sentences. Use bullet points.
> - Don't write everything the speaker says. Write key facts, concepts and numbers.
> - Use abbreviations.
> - Read your notes as soon as possible after writing them and make sure you can remember the necessary information.
> - Write up your notes for future reference if necessary.

Spider Silk

Speaking

1 Choose a material, traditional or new, which you know something about. Under the headings *Characteristics/properties* and *Applications/uses*, make notes in your notebook for a short presentation.

2 Practise your presentation with a partner. Check any pronunciation in a dictionary or with your teacher.

3 Present your material to the group. The group must listen and refer to the mechanical properties you have studied in Lessons 3 and 4, and ask questions about any of them which are not included in the talk.

➡ Workbook pages 6/7

Speaking and vocabulary

In 2005, a competition was held by Southampton University in Britain for young people aged seven to 19 to design an aircraft for the year 2050. Below are three of the designs which won prizes.

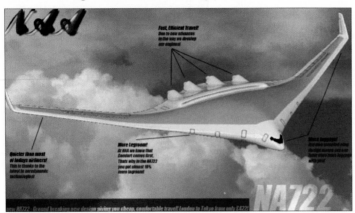

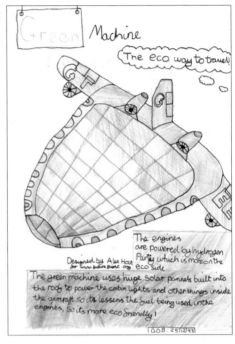

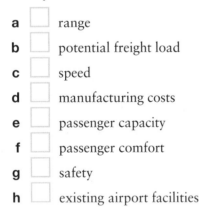

1 Look at the pictures and discuss how much information they give. Which picture gives the best general idea about the plane? Why?

2 What are the most important constraints on aircraft design? Number the following in order of importance.

a ☐ range

b ☐ potential freight load

c ☐ speed

d ☐ manufacturing costs

e ☐ passenger capacity

f ☐ passenger comfort

g ☐ safety

h ☐ existing airport facilities

Reading and vocabulary

1 Check the meaning of the words in **bold** in the following text and record them in your vocabulary log.

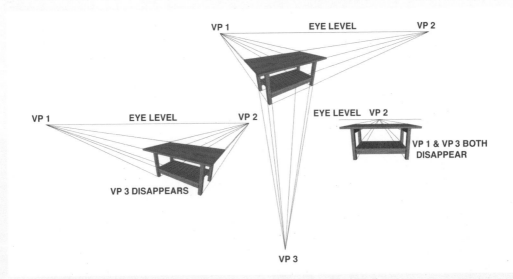

When something new is designed, it is necessary to produce detailed technical drawings by **orthographic projection** so that it can be built correctly. The problem with these drawings is that they are **two-dimensional** but the world around us is **three-dimensional,** and it is sometimes difficult to **visualise** what the finished product
5   will look like. Because of this, designers often produce **pictorial drawings** as well as orthographic ones. Sometimes a quick rough sketch is enough, but often this doesn't provide enough information. A better way of producing a pictorial view is to use **perspective**. The history of this drawing **technique** goes back a thousand years to the ideas and discoveries of the Arab mathematician and philosopher, Ibn al-Haitham
10 (965–1040). It was his work in **geometry** and **optics** that enabled later artists and engineers to develop the skill of drawing things in a realistic way. There are a number of different ways of drawing perspective, but they all give the **impression** that you are actually looking at the object from a specific **point of view**. The most common type of perspective is called **two-point perspective** because it is based on two
15 vanishing points which are situated on a horizontal line known as **eye-level**. There is a more complex kind of perspective called **three-point perspective** which gives an even more realistic impression.

2 Read the passage again and label the different types of pictures on this page.

3 With a partner, discuss the advantages of perspective projection, orthographic projection and pictorial drawings.

⇨ Workbook pages 7/8

Back to the future

Vocabulary

1 **Look at the picture below and discuss the questions with a partner.**

   a What kind of machine is it?

   b How old is it?

2 **Here are some facts and figures about one modern airship, the Zeppelin-NT. Read them and underline anything you find surprising or interesting.**

   a filled with: helium gas

   b length: 75 m

   c maximum width: 19.4 m

   d height: 17.4 m

   e volume of main body: 10,000+ m^3

   f cabin volume: 26 m^3

   g cabin length: 10.7 m

   h no. of crew: 2

   i passenger capacity: 12

   j max. take-off weight: 10,690 kg

   k payload: 1,900 kg

   l power plant: 3 x 200 hp Textron Lycoming 10-360 engines

   m max. speed: 125 km/h

   n range: 900 km

   o ceiling: 2,600 m

   p max. flight duration: +/– 24 hours

3 What units do the following refer to?

a m _____

b kg _____

c km/h _____

d hp _____

e m³ _____

4 What are these mathematical symbols?

a +/– _____

b % _____

c < > _____

d ÷ _____

e x _____

f ¼ , ½ , ¾ _____

5 Write some more units and symbols that are used in your technical field.

6 How would you calculate:

a the circumference of a circle?

b the surface area of a sphere?

c 30 m/h in km/h?

d the volume of a cube?

Language and speaking

1 Form questions about each of the 16 specifications in Vocabulary Exercise 2. Try to think of questions which start with something other than *What is ...* For example, for i) passenger capacity:

What is the passenger capacity?

or

How many passengers can it carry?

2 Work in pairs, A and B. Student A should look at the information sheet on page 242. Student B should look at page 243. Find out information about your partner's aircraft using the questions you came up with in Exercise 1.

3 Some people think that the airliners of the future will be like airships and not like modern aircraft at all. What do you think? Discuss how you think air travel will change in the future. Use the language in the Language Box.

➡ Workbook page 9

Language Box

Predictions

To talk about ideas and suggested projects in the future, you can often use expressions with *will* + adverb:

People will probably travel more.

Fuel will definitely be non-carbon based.

Airports will certainly need to be bigger.

Design rationale

Vocabulary

1 **Find seven words connected with aeroplanes and flights in the word search.**

D	R	A	I	S	P	P	S	I	D
I	A	D	D	E	N	G	I	N	E
S	D	I	M	E	N	D	S	M	E
R	A	T	R	R	C	S	L	I	I
A	T	A	I	B	D	O	R	S	D
S	R	A	D	A	R	S	Y	S	T
F	D	I	S	P	L	A	Y	I	N
S	S	D	F	T	A	N	K	L	G
F	O	R	E	P	L	A	N	E	E

2 **Now look at this diagram of a jet fighter. (The diagram is called a *cutaway* because part of the outside has been cut away to show the inside.)**

a Finish labelling the cutaway using six words from the word search.

b Use a dictionary or the glossary to make sure you know the meaning of the compound words and phrases.

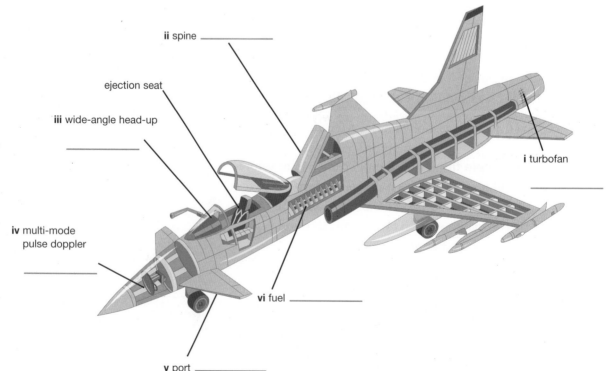

ii spine _____

ejection seat

iii wide-angle head-up
_____

i turbofan

iv multi-mode
pulse doppler
_____

vi fuel _____

v port _____

3 **Where do you think the following labels should go on the cutaway?**

retractable flight refuelling probe

external fuel tank

medium-range active missile

4 **Work with a partner and look at the words below.**

a Use the glossary or a dictionary to find another word that can go before or after each word to make a compound, for example: *turbofan engine, flight plan.*

b Compare your ideas with another pair and explain any new expressions to each other.

i	wing	_____	**ix**	steel	_____
ii	fuselage	_____	**x**	design	_____
iii	tail	_____	**xi**	materials	_____
iv	surface	_____	**xii**	component	_____
v	fuel	_____	**xiii**	costs	_____
vi	engine	*turbofan engine*	**xiv**	capacity	_____
vii	port/starboard	_____	**xv**	air	_____
viii	flight	*flight plan*			

Speaking and writing

1 **Answer these questions.**

a Why do you think the jet fighter carries short-range missiles under the outside section of the wing and medium-range missiles inside them?

b What is the exact purpose of the foreplanes?

2 **Read the Language Box. With a partner, take turns to choose a labelled part of the cutaway and ask about its function. Your partner will respond using some of the expressions in the box.**
Example:

> What's the airbrake for?

> It's for slowing the plane down. When the plane lands, it lifts up. It's designed like this to increase drag.

Language Box

Reason and purpose

These expressions will be useful when you need to explain ideas:

X is for ~ing.
What this/X does is …
The reason (for this) is that …
That's because …

X + verb:
X is designed like this so that …

Question forms

Most questions in English are formed using the rule 'auxiliary + subject + verb':
Will it break?
Does heat pass through it?

The verb *be* is different:
Is it suitable?
How heavy is it?

Passive questions use *be* + past participle:
Can the surface be damaged easily?

⇨Workbook pages 10/11

White elephants?

Vocabulary

1 How many different reasons can you think of why an aircraft design project could fail?

2 Work with a partner. Look at the list below of reasons why projects are sometimes abandoned.

a Check that you understand and can pronounce all of the phrases.

b Make a table in your notebook with the headings: *Politics, Design, Economics*.

c Group the reasons under the headings in the table.

d Can you add any more to the lists?

- negative public opinion
- need for investment in new infrastructure, e.g., runways, hangars, machinery
- withdrawal of funding
- accidents during trials
- delays in production

- innovative production techniques are expensive
- high manufacturing costs
- lack of interest from potential buyers
- changing commercial or military needs
- competition from other companies

Politics	Design	Economics

Reading

1 You are going to read about an aircraft project which failed. Look at the text opposite, but do not read it yet.

a From the picture and your own knowledge, what do you think the text will say?

b Read the text quickly to see how many of your ideas are there.

2 Read the text again more slowly and correct the information in these sentences.

a The American government put US$7 million into the project.

b The total cost of the project was US$18 million.

c The plane spent 33 years in a museum.

d The plane flew several times.

e Hughes designed the plane in 1924.

Skills Box 1

Predicting content

You can make it easier to read a text by thinking: *What can I guess about this topic? What do I already know?*

Look at the titles, pictures and layout to help you.

Skills Box 2

Reading for specific information

As well as predicting, you need to be able to understand the main details of the text. You do not need to worry about understanding every word, but you should be able to pick out key information in a longer text.

The Spruce Goose

The journey from the initial idea and design specifications to a finished aircraft is often long and complex. Sometimes it is so long and so complex
5   that projects fail. They may fail at any stage, in some cases not even getting off the drawing board. Sometimes there are serious technical problems with the prototype which cannot be easily solved.

Sometimes the cost of production proves to be much too high and the project is
10   scrapped for financial reasons, even after millions of dollars have been spent. Sometimes, by the time a prototype has been produced, there is no longer any need for the aircraft.

One example of such a *white elephant* is the HK1 military transport, which became known as the 'Spruce Goose'. In 1942, during the Second World War, the millionaire
15   businessman and engineer Howard Hughes took on responsibility for designing and building the largest plane in the world. The design brief for this enormous plane was very demanding. The plane had to transport men and materials in very large quantities across the Atlantic without stopping to refuel. It had to be able to land on water as well as land. Because it was wartime, aluminium and steel were in short supply, so most of
20   the aircraft was built of alternative materials.

Because this was such an innovative project, many years were spent on research and development. This cost a lot of money. By the end of the war, the American government had invested $18 million in the project and decided that the aircraft was too expensive and that they no longer needed it. However, Hughes was determined to finish the
25   project and put in $7 million of his own money. He completed a prototype and flew the 'Goose' for its first and only flight. Although the flight was a success, the American government didn't change its mind and the plane spent the next 33 years 'ready for flight' in a storage hangar at a cost of $1 million dollars a year. Finally, after Hughes died, the plane was sent to a museum.

Language

1   **Look at this sentence: 'Sometimes there are technical problems which cannot *be easily solved*.'**

Why is the passive form used here instead of the other option: 'Sometimes there are technical problems which the team cannot solve'?

2   **Find further examples of the use of passive forms in the text about the Spruce Goose: three in paragraph 1; one in paragraph 2 and two in paragraph 3.**

➡ **Workbook pages 12/13**

Lesson 9
Lost classics

Reading

Here are two more aircraft that never went beyond a prototype. What is unusual about them? Study the drawings and specifications and suggest reasons why you think these planes were never produced.

Bristol *Brabazon* Airliner

Custer short takeoff Channel Wing

Manufacturer	Bristol Aeroplane Company	
Dimensions		
Length	177 ft	53.95 m
Wingspan	230 ft	70.1 m
Height	50 ft	15.24 m
Wing area	5,317 ft²	493.95 m²
Weights		
Empty	145,100 lb	65,816 kg
Maximum take-off	290,000 lb	13,542 kg
Powerplant		
Engine	8 Bristol Centauras	
Power	2,650 hp	1,864 kW
Performance		
Maximum speed	330 m/h	483 km/h
Cruise speed	250 m/h	402 km/h
Range	5,500 miles	8,850 km
Ceiling	24,500 ft	7,620 m
No. of passengers	50–80	

Manufacturer	Custer Channel Wing Corporation	
Dimensions		
Length	28 ft 8 in	8.76 m
Wingspan	41 ft 2 in	12.55 m
Height at cabin	6 ft 9 in	2.05 m
Weights		
Empty	3,000 lb	1,360 kg
Maximum take-off	5,400 lb	2,449 kg
Powerplant		
Engine	2x Continental IO-470D	
Power	260 hp	193.89 kW
Performance		
Maximum speed	200 m/h	331 km/h
Cruise speed	180 m/h	289.6 km/h
Range	1,150 miles	1,850 km
Ceiling	20,000 ft	6,096 m
Take-off distance	<200 ft	<61 m
No. of passengers	5	

Vocabulary

The data about these planes is given in metric and imperial measurements. Imperial measurements are still used in some parts of the aircraft industry. Can you estimate the metric or imperial equivalents of the quantities below? Use the information in the tables in the Reading to make a rough estimate first, then check your answer on a calculator.

a 1,000 ft is approximately _____ m.

b 500 kg is approximately _____ lb.

c 800 km/h is approximately _____ m/h.

d 90 kW is approximately _____ hp.

e 100 m² is approximately _____ ft².

f 6 in is approximately _____ cm.

Language

1 **Look at the following sentences. Underline the time expression in each.**

a By the end of 1969, the second prototype was ready for trials.

b Until the 1970s, passenger airliners had always been considered like ocean liners.

c The government decided that it no longer wished to invest in warplanes.

d Finally, a specially re-formed team completed the project a year later.

e After three years' work and US$9 million in investment, the plug was pulled.

f While trials continued, plans were being made for a third prototype.

g Unfortunately, funding was then frozen for the next 12 months.

h Initially, there was great support for the idea.

i From that moment on, the problems multiplied.

j They had been considered the biggest potential customer since the beginning.

2 **Look at the Language Box. What tense does the verb in each sentence in Exercise 1 refer to? Mark them PS (past simple), PC (past continuous) or PP (past perfect).**

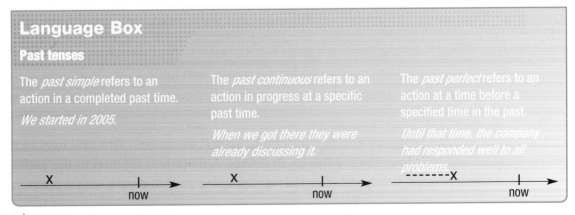

Language Box

Past tenses

The *past simple* refers to an action in a completed past time.

We started in 2005.

The *past continuous* refers to an action in progress at a specific past time.

When we got there they were already discussing it.

The *past perfect* refers to an action at a time before a specified time in the past.

Until that time, the company had responded well to all problems.

⇨Workbook pages 13/14

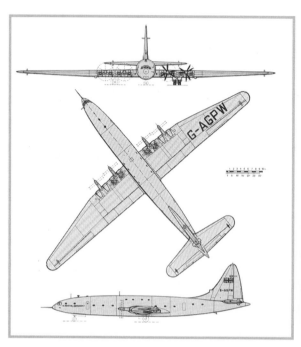

Vocabulary

1 **Complete the sentences below with words from the box.**

> cabin   production   speeds   take-off   hangar   prototypes   wingspan   passengers

a It was the largest plane ever built in Britain and had a _____ of 70 metres.

b It was designed to carry a maximum of just 100 _____.

c There was a lot of vibration, which made the _____ noisy.

d It could fly at very low _____ because of its unusual design.

e The great _____ that was built for it is now used for the Airbus A380.

f It never became a commercial aircraft, although several _____ were built.

g It flew several times at air shows but never went into full _____.

h Its eight powerful engines allowed it to lift and _____ quite easily.

2 **Look at each sentence and decide which plane it describes. Write (B) next to the sentences that describe the Bristol Brabazon Airliner and (C) next to the sentences that describe the Custer Channel Wing. Compare your answers with a partner.**

Writing

1   You are going to write a brief history of one of the projects in Lesson 9 – either the Custer Channel Wing or the Bristol Brabazon Airliner. Choose whether you want to:

    a use your own ideas to create a probable background to the project.

    or

    b do some research on the Internet to find out the real story.

2   Copy this table into your notebook. Work with a partner and put each of the phrases in the list below under one of the headings.

need for a new aircraft	design brief	R&D	reasons for failure	specifications

Unfortunately,

length; weight (empty)

There was no existing aircraft which

It had become clear that

These requirements were very demanding

A prototype was completed

Traditional materials were not suitable

Much time was spent on

It had to be able to

3   Talk to a partner. Can you add any other grammar and vocabulary from the unit?

4   Write your text. Use the same subheadings as above. When you have finished, compare texts with your colleagues.

Speaking

Work with a partner. Choose one of the photos above. Describe it in terms of cost, appearance, durability, applications, etc. Your partner should guess which aeroplane you are describing.

➪Workbook pages 14–16

Although modern manufacturing is dependent on powered machinery, there are still some jobs that can only be done by hand. Even in the most modern and well-equipped factory or workshop, you will find a wide range of tools that require the skill, knowledge and muscles of the person using them.

Vocabulary

1 **Look at the words in the box below.**

> clean   jaws   tighten   calibrate   teeth   oil   blade
> head   edge   sharpen   handle   hit   put   cover

a Put these words into two groups, nouns and verbs (some words may go in both groups).

nouns	verbs

b Which words go with the tools in these pictures?

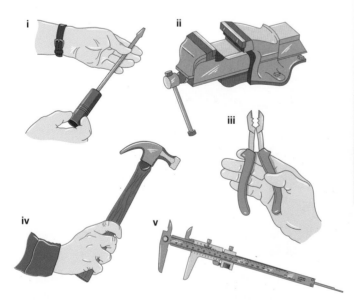

Language Box 1

Imperatives

In English, sentences usually contain a subject followed by a verb, e.g.:

He sharpened the blade.

Imperative sentences giving instructions, orders or warnings have a verb but no subject:

Put those tools away.
Don't cut that wire!

c Look at Language Box 1. Make imperative sentences using the verbs and tools from this exercise.

2 Hand tools are used for a variety of purposes. Check the meaning of these verbs with a partner or in your dictionary. Can you think of others?

> measuring   gripping   bending   cutting   pounding   punching   screwing   threading

3 Match each of the verbs with at least one of the tools in Vocabulary Exercise 1.

4 Check that you know the pronunciation of all of the vocabulary in Exercises 1 to 3.

Speaking

There are three golden rules for using hand tools.

1 Use them for their correct purpose.

2 Make sure they are in good condition.

3 Keep them in their proper place.

> ### Language Box 2
>
> **Preposition + verb + ~ing**
>
> Prepositions such as *for, in, from* and *by* are followed by the *~ing* form of a verb or by a noun phrase:
> It's designed *for gripping.*
> This stops it *from rusting.*
> It's used *in metal polishing.*
> You start *by heating it.*

1 Look at Language Box 2. In pairs, discuss the three rules by answering these questions.

a What is the correct purpose of each tool in the pictures?

b How do you keep them in good condition?

c Where should they be kept?

2 Look at the description in the table below. Guess what type of tool is being described.

components	handle, frame and removable blade
size	small (junior variety) to large (similar to band saw)
uses	used for cutting a variety of metals
used by	engineers, plumbers, etc.
power	hand or mechanical tool
safety	sharp blade, may fracture

3 Work with a partner. Choose two more tools and complete the tables.

components	
size	
uses	
used by	
power	
safety	

components	
size	
uses	
used by	
power	
safety	

⇨ Workbook page 17

Blade manufacture

Vocabulary and reading

1 Look at the two lists of words and make possible combinations using one word from each list, e.g., *superb quality*.

superb	shape
original	quality
long	shape
national	polisher
complex	sharp
rough	history
specialist	process
razor	treasure

2 Read the following text about sword manufacture as fast as you can and circle the combinations from Exercise 1.

More than a thousand years ago, the great Arab city of Damascus was not only the centre of the Middle Eastern sword trade, but was also famous for the superb quality of the swords manufactured in the city. It was said that a Damascene sword blade was so flexible that it could be bent around a man's body and then return to its original
5 shape without any damage.
Unfortunately, nobody knows exactly how these wonderful swords were made.

Another country which has a long history of sword-making is Japan, where there are
10 still a few craftsmen who make samurai katana swords in the traditional way. The most famous of these is Sumitani Masamine, whose skill is so highly respected that he has been given the title of 'National Living Treasure' by the Japanese government. Traditionally, the production of a katana sword was a complex process, which often involved several craftsmen; there was a smith to forge the sword into a rough shape,
15 another smith to form and work the metal, a specialist polisher, and finally, another specialist to form the razor sharp edge.

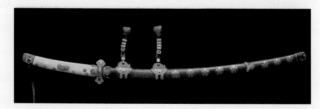

3 Complete the gaps in the next part of the text with the correct form of the words in the box.

> cool   beat   tie   fill   heat   hammer

It is believed that Damascene craftsmen started by _____ a batch of low-carbon wrought iron into thin sheets. These were then _____ tightly into bundles. Next, high-carbon cast iron was _____ until it was molten, and then the bundles of wrought iron were thrown in. The air spaces between the sheets

5 were _____ with the molten metal, which had the effect of welding the bundles into a solid lump of metal. This was then _____ into the rough shape of a sword while it was still hot. Finally, after _____, the blade was filed, ground and polished.

4 Check that you know the pronunciation of the words you used to fill the gaps in Exercise 3.

Listening

You are going to listen to part of a radio interview with a collector of Japanese swords.

1 🎧 Listen once and make a mark in your notebook each time you hear one of the words in Vocabulary and reading Exercise 3.

> ### Skills Box
> **Listening for key items**
> You will need to practise identifying key items in a stream of speech. These could be numbers, names or ideas.

2 🎧 Listen again. Complete the information about the manufacturing process.

   a overall manufacturing style: heating and _____

   b process: swordmaker heats, folds and _____ metal to eliminate _____

   c blade: _____ not very flexible, only one _____

3 In pairs, talk about the differences between the two manufacturing processes.

➥Workbook pages 18/19

Speaking

1 What sort of machine tools are you familiar with? Make a list and compare it with a partner.

2 The machines below are computer numerical control (CNC) machine tools. In pairs, discuss the following questions.

a What do you think *computer numerical control* means?

b Have you ever seen or used CNC tools?

c What advantages do they have over hand-operated machines?

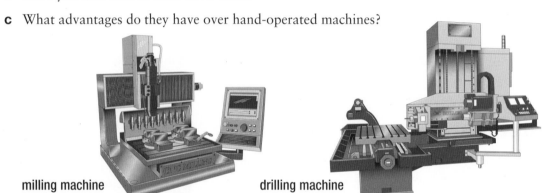

milling machine   drilling machine

Vocabulary

Look at the following words in the introduction to a description of CNC tools. The word *production* occurs as *mass-production* and *speed of production*, two very common combinations. In what combinations do the other three words occur?

production (lines 2 and 10)   equipment (line 8)   machines (line 6)   main (lines 7 and 9)

CNC machine tools

Mass-production of large aircraft, both commercial and military, would be impossible without the use of computer numerical
5 control (CNC) machine tools. Although hand-operated machines and tools are still used, most of the main parts of the airframe are now produced by computerised equipment. There are three main reasons for this: speed
10 of production, accuracy and repeatability.

> **Vocabulary Box**
>
> **Word combinations**
>
> Vocabulary always occurs as combinations of words, so memorise the whole phrase, not just a single word.
>
> This system will help to improve your reading, writing, listening and speaking skills.

Reading and speaking

1. Circle these numbers or quantities in the text below: *5, 3, 70, 1,000.*

2. Look again. What do they refer to?

3. Read the text again more carefully and transfer the information it gives into the table.

	basic function	problem in the past	advantage of this system
ICY			
FAM			
drilling machine			

In the past, parts were often tailored specially to fit one particular aircraft and could not be fitted to another of the same design. This was because the tolerances and margins of error were too wide.

Nowadays, however, manufacturers try to make parts interchangeable by using ICY
5 (an industry abbreviation for interchangeability). At the Eurofighter component factory in Samlesbury, some aircraft parts are machined to tolerances as fine as 70 microns. The tolerance is the difference between the maximum and minimum acceptable dimension: the upper and lower limits. Making components to this very fine tolerance means that if there is a hold-up with one plane, a part can be fitted to
10 another one without modification. In this way, the production process is kept moving.

One of the key types of machine at this factory is the five-axis milling machine (FAM). Milling machines are used for forming surfaces and are extremely versatile. By the use of different configurations and different-shaped cutters, a wide variety of shapes and surfaces can be machined. Before this, standard milling machines were designed to
15 work on only three axes, which limits the amount of cutting that can be done at any one time without changing the position of the workpiece. The FAM machine can operate along five separate axes and is able to produce highly complex curved shapes and surfaces in a short time.

Another type of CNC machine being used at the Samlesbury factory is a drilling
20 machine. Traditionally, holes were drilled in aircraft skins by hand. However, there is now a computer-controlled drill that can automatically make a thousand holes in the front fuselage of the Eurofighter. This speeds up production and produces much more accurate load-bearing panels. Some modern CNC drills have multiple spindles, which enable several drilling operations to be carried out on a workpiece simultaneously.

4. Work in groups and give a short talk about one of the machine tools in the table.
⇨ Workbook pages 19/20

Two different drills

Speaking

Look at these two drills and talk about the differences between them: talk about what they are made of, how they work and what they are used for. What is the reason for the different shapes and weights?

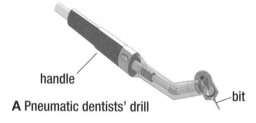

handle

A Pneumatic dentists' drill

bit

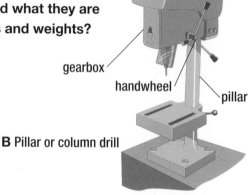

gearbox

handwheel

pillar

B Pillar or column drill

Listening

1   Discuss the following notes about the drills with a partner. Which of them do you think are about A, which are about B and which are about both? Tick (✓) the columns.

		A	B	Both
a	drill bit on more than one axis			
b	plastic handle			
c	power transmitted by gears			
d	base helps prevent movement			
e	pillar supports worktable			
f	air-powered			
g	unsuitable for teeth			
h	bit can be accurately positioned			
i	position of bit controlled by handwheel			

2   🎧 Listen and check your answers to Exercise 1.

3   Try to expand the words and phrases in the table to make full sentences.

4   🎧 Listen and compare your ideas with the sentences on the recording.

5   Look at the Skills Box opposite. In the tapescript, mark stressed words on the sentences from Exercise 4.

Vocabulary

1. Match these words with the correct definition:
adjust (*v*); align (*v*); spanner (*n*); workpiece (*n*); turning tool (*n*).

a _____ : to arrange in a line or certain direction

b _____ : a piece of metal that is to be machined

c _____ : a tool for shaping wood or metal

d _____ : to change to suit new conditions

e _____ : a tool for gripping and twisting objects

2. In pairs, write an example sentence for each word.

Language

1. Drilling machines can be extremely dangerous if they are not used correctly. Look at this list of safety rules. Match each sentence beginning on the left to an ending on the right so that they make sense.

a	Don't adjust the machine	away from turning tools.
b	Never clean up	before operating the machine.
c	All loose clothing should be kept	on the drilling table.
d	Make sure that cutting tools are correctly aligned	while operating a drilling machine.
e	Remove all chuck keys and spanners	while the machine is operating.
f	Always wear eye protection	to avoid damaged tools and workpieces.
g	Never put tools or equipment	with your hands. Use a brush.
h	Slow down the feed as the drill breaks through the work	before starting the machine.

2. Write similar rules for using a machine tool that you are familiar with.

⇨ Workbook pages 20/21

Lesson 5
The history of the lathe

Speaking

1 How have the following discoveries and inventions made life easier? Discuss your ideas with a partner.

> irrigation   penicillin   the internal combustion engine
> the gyroscope   writing   the microprocessor

2 What other discoveries and inventions have helped to improve people's lives?

Reading and vocabulary

1 What technology is shown in all four pictures below? Check your ideas with a partner.

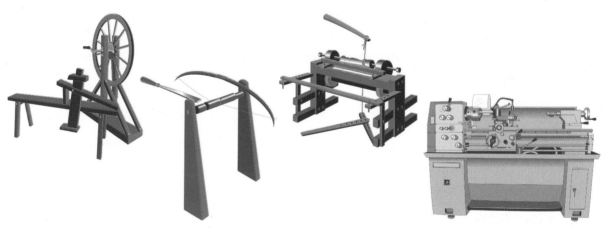

2 Read the following paragraph quickly to see if you were right.

> How long has man been turning wood? Almost certainly longer than we have evidence for! What did the first lathe look like? We are not sure, but we can come to a reasonable conclusion, bearing in mind the materials and technology available. There are just a few early illustrations that give us some insight, plus the continuing use of simple technology in many parts of the world.

3 Can you name some modern products that are 'turned' during the production process?

4 Now read the rest of the text on the opposite page. Label the pictures in Exercise 1 with the bold words from the text.

One thing is certain: all early lathes would have been of the reciprocal variety; that is to say, the material to be turned would have been supported between two centres and spun backwards and forwards in some way, with a **strap**, a **bow** or
5 a **pole**. Many people will be familiar with this concept via the 'pole lathe', as it is still used today by certain traditional chair makers, both amateur and professional.

It is a drawing, or rather a simple sketch, by the Italian genius Leonardo da Vinci, c.1480, that affords us our first glimpse of
10 what an early treadle wheel lathe looked like. The main elements required for foot-propelled continuous rotation are clearly shown for the first time; the **flywheel, crank** and **treadle**. It was the crank, in conjunction with the flywheel, that provided a huge technological advance (this is still used in our modern internal combustion engines). The crank, linked to a treadle, provided constant rotation, while the momentum of the large
15 flywheel ensured the crank was carried over its 'dead spot'.

Whether da Vinci actually designed this treadle lathe or whether he just sketched what was already in existence will always be a matter of debate, but there is no doubt that without it there would not have been – could not have been – an Industrial Revolution! This lathe is the first machine tool, the father of all others, which went on to produce increasingly
20 complex machines, leading to the Industrial Age of the 19th and 20th centuries.

Language

1 **Look again at the text in the previous section. Complete the chart with words from the box.**

> reciprocal   material   treadle wheel   flywheel   crank
> more complex   bow   pole   ~~strap~~   rotation   ~~treadle~~

a Early lathes:

_____ variety:
worked by spinning
_____ between
two centres and used:

i _____

ii _____

iii _strap_

b The _____
lathe:
worked by foot-propelled
continuous _____
and included:

i _treadle_

ii _____

iii _____

c _____
machines:
partly responsible for the
_____

2 **With a partner, use the following expressions to describe the operation of a lathe to a person who is not familiar with it.**

> It consists of   It is powered by   X is connected to Y, which
> This in turn   the A turns/moves/drives the B

⇨**Workbook pages 21/22**

Lesson 6
Modern lathes

Reading and speaking

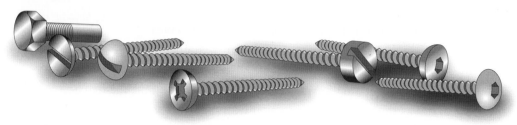

1   **Look at the screws and bolts in the pictures. With a partner, discuss whether the following statements are true (T) or false (F).**

   a ____ Screws are nearly always made from steel.

   b ____ Screw heads come in many shapes and sizes.

   c ____ All screws need to be turned clockwise when they are tightened.

   d ____ The dimensions of most screws and bolts are standardised.

   e ____ The same system of standardisation is used all over the world.

2   **Read the passage and check your answers.**

   Screws and bolts are made in a wide range of materials, including brass, bronze, aluminium and titanium as well as steel. They consist of a shaft with a thread and a head that may come in a variety of shapes, including round, flat, oval, button or cap. The vast majority of screws have a right-hand thread, which means they are tightened
5 by clockwise rotation. Occasionally, screws have a left-hand thread, such as the left-hand pedal on a bicycle.

   Machine screws come in a range of sizes, and around 85% of all screws and bolts are designed to unified thread dimensions. They are categorised according to their diameter and pitch, or number of threads per inch. ISO metric screw thread is the
10 most popular standard and has displaced previous systems, but there are other common systems, including the British Standard Whitworth, BA system (British Association) and the SAE Unified Thread Standard.

3   **Discuss the following questions in groups.**

   a Why is it important to have standardised screws and bolts in the modern world?

   b Which industries rely on mass-production of fasteners?

   c When did screws and bolts start to become standardised?

   d What machinery is needed to produce standardised screws and bolts?

Listening

You are going to listen to someone talking about the development of the modern lathe.

1 🎧 **Listen and tick the questions from Reading and speaking Exercise 3 that the man answers.**

2 **Complete the text below with words from the box.**

> gearbox   workpiece   rotational   threads   lead screw   rotational

The **a**_____ is a long, threaded rod that carries a tool along the axis of a rotating workpiece. It ensures that the **b**_____ moves at a constant, even speed so that **c**_____ can be cut into it. The relationship between the **d**_____ speed of the tool and the **e**_____ speed of the workpiece can be varied by means of a **f**_____.

3 🎧 **Listen and check your answers.**

Vocabulary

The diagram shows a schematic view of a modern lathe with some of its main parts.

1 **Read the descriptions of the parts and match them with the correct number.**

a _____ BED: the main body of the lathe.

b _____ LEAD SCREW: the long, threaded rod which controls the longitudinal speed of the saddle.

c _____ HEADSTOCK: the box, always to the left of the operator, containing the gears which control the spindle speed.

d _____ TAILSTOCK: the moveable block at the right-hand end of the lathe which is used to support the centre or to hold drills.

e _____ SPINDLE SPEED SELECTOR: the knob or lever used to control the rotational speed (rpm) of the spindle.

f _____ SADDLE: the metal block which supports the tool and moves longitudinally between the headstock and the tailstock.

g _____ SPINDLE: the rotating shaft to which the workpiece is attached.

2 **With a partner, take it in turns to ask each other questions about the different parts of the lathe.**

⇨**Workbook pages 22/23**

Lesson 7
Aluminium

Reading

1 **Read the text below quickly and decide which of these three titles you think is best.**

  a A comparison of industrial metals

  b Useful properties of aluminium

  c Industrial chemicals

Since the Wright brothers chose an aluminium engine for their historic first flight in 1903, this versatile metal has become increasingly important in the aircraft industry. Today, aluminium comprises 70 to 80 per cent of the weight of modern commercial aircraft. Why is it so important?

5 Probably the main reason is that it is a strong and very light metal with a density which is about a third that of steel. Pure aluminium weighs only 2.7 g per cubic centimetre, compared with 7.8 g for mild steel. This reduces the dead weight of an aircraft and the energy consumption, making it possible to carry heavy loads such as passengers or cargo.

10 Secondly, it is highly corrosion-resistant. When it is exposed to the atmosphere, a thin invisible skin of oxide forms but, unlike steel, this process of oxidisation does not eat any further into the metal. The layer of oxide protects the metal from further corrosion, and this enables the metal to last a long time.

Because it is a relatively soft metal, aluminium can be cut, drilled, machined and 15 formed into different shapes much more easily than steel. This is extremely important for manufacturing planes quickly and cheaply. In its pure form, however, aluminium is too soft and too ductile to be suitable, so alloys containing metals such as copper, manganese, chromium and zinc are produced. Different alloys are used for the different parts of the plane, depending on the strength and other properties required.

20 Because they are not pure aluminium, these alloys are often susceptible to corrosion. To avoid this, alloy sheets for aircraft are sometimes electrochemically coated with pure aluminium. As long as this protective coating stays intact, the underlying alloy is unaffected by harmful substances and the durability of the metal is increased.

In addition to its mechanical properties and corrosion resistance, aluminium has high 25 conductivity, both electrical and thermal. Although less effective as a conductor than copper, it is cheaper and as a result is the most commonly used material in main electrical power lines.

2 **Now read the text more carefully. Which set of notes, A or B below, represents the information better? Why?**

A

Since 1903 (Wrights), Al v. imp: 70-80% weight of modern plane. Why?

1 <u>Strong+light.</u> Al 2.7g/cm³, mild steel 7.8g/cm³, so less fuel used, more cargo capacity with Al

2 <u>Resists corrosion</u>, long-lasting: Al oxide forms in air; protects metal from corrosion

3 <u>Soft + easier</u> to work than steel, so plane production quick + cheap. Alloys with copper, manganese, chromium and zinc used for extra strength/ other properties. But alloys not pure Al, so corrosion a problem - sometimes coated with pure Al for durability

4 <u>Good conductor</u> of elec. (and heat), - used in power lines.

B

· Al important since 1903 – Wrights' engine made of aluminium. Used in modern planes. Why?

· Al is strong and light. Density is a third of steel. Al weighs 2.7g/cm³, mild steel weighs 7.8g/cm³. This is very important in designing aircraft. More passengers can be carried.

· Secondly, it is highly corrosion-resistant. Unlike in the case of steel, this process of oxidisation does not eat any further into the metal. The layer of oxide protects the metal. This enables the metal to last a long time.

· Aluminium can be cut, drilled, machined and formed into different shapes. This is extremely important for manufacturing planes quickly and cheaply. In its pure form, however, aluminium alloys containing copper, manganese, chromium and zinc are produced. Harder and stronger.

· Alloys are more susceptible to corrosion.

· In addition to its mechanical properties and corrosion resistance, aluminium has high conductivity and is used in electrical power lines.

3 **Circle the following in the text in Reading. What do they refer to?**

this (line 2)   this (line 7)   it (line 10)   its (line 16)   these (line 20)   this (line 21)   it (line 26)

Vocabulary and speaking

1 **In the set of notes A above, what do the following abbreviations mean?**

e.g.   v. imp   +   Al   elec.   g/cm³

2 **Find the expressions from the box below in the text and look at how they are used. Use them to fill in the gaps in a to e.**

> However   secondly   Because   so   In addition to

a _____ gold does not rust easily, it is often used in jewellery and dentistry.

b Asbestos is a good insulator. _____ , the dust is highly toxic.

c There are two reasons: firstly, water is cheaper than chemical options; _____ , it is easier to store.

d _____ the high cost of raw materials, other problems include the long development process.

e Common cutting tools need to be rust-free, _____ they are often made of stainless steel.

⇨**Workbook pages 23/24**

Working with Alclad

Vocabulary

Complete the table with the correct forms of the words. Check that you know the correct pronunciation of all the words.

adjectives	nouns
soft	
malleable	
ductile	
	lightness
	resistance
conductive	
	density
durable	
	flexibility
tough	

Speaking and writing

1 In pairs, talk about the properties of aluminium which make it suitable for these products.

2 Write a sentence for each of the objects in the pictures. For example:
   Aluminium is good for making windows because it is tough and durable.

Listening

1 Before you listen, check that you know what these words mean.

> bend   shears   scriber   mark out   edge   smooth   handle   hole   radius   burr

2 🎧 Listen to an instructor in a workshop talking to some students about working with a material called Alclad. How many of the following does he tell them how to do: *cutting*, *drilling*, *marking out*, *bending*?

3 🎧 Listen again. Decide whether the statements below are true (T) or false (F).

a ___ The Alcladding process stops corrosion.

b ___ You have to handle all aluminium sheets very carefully.

c ___ There are five main operations you do on sheet metal.

d ___ You mustn't use a scriber on Alcladded metal sheets.

e ___ You must make sure the edges are completely smooth.

f ___ You mustn't bend the sheets too much.

4 🎧 Listen again to the instructor and follow in the tapescript. Choose five or six expressions from the text which you think would be useful to remember for asking about processes and giving instructions, for example:

- Why do you …?
- What do you do …?
- You have to …
- It's really important to …

Speaking

Write and rehearse a short conversation in which the instructor either:

a comments on John's and Martin's work on the Alclad sheet;

or

b instructs the two young men to make some additional bends or cuts in the sheet.

Include the language you identified in Listening Exercise 4.

Check that you know the correct pronunciation of all the words.

➡ Workbook pages 24/25

Lesson 9
Fastenings

Speaking

1 Match each of these words to one of the pictures below: *screw, bolt, rivet, pin*.
How are these fasteners different from each other?

 a   b    c   d

2 How are these fasteners used in everyday life?

Reading

1 Look at the diagrams below and read the explanations of the 'standard' and 'blind' riveting processes. Which process do the diagrams show?

A 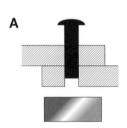   B    C

a In standard riveting:

  i a hole is drilled through the sheets to be joined and the rivet is inserted;

  ii the rivet is pushed or driven in with a hammer from one side against a bar on the other;

  iii the bar acts as a resistance, which expands the other end of the rivet so that it holds the sheets together.

b A blind rivet consists of two parts: the rivet body and a pin which goes inside it.

  i The blind rivet is inserted into the pre-drilled hole.

  ii The riveting tool is placed over the pin. When the tool is actuated, the jaws of the tool grip the pin.

  iii The pin head is pulled into the rivet body with great force, expanding the end of the rivet.

  iv The pin then breaks off, leaving the rivet holding the pinhead.

2   **Label the diagrams in Exercise 1 with the words in the box.**

hammer   bar   sheet

3   **Now read about riveting in the aircraft industry. Write in the missing sentences *a* to *c* which start each paragraph *i* to *iii*.**

a   Blind riveting was first devised so that sheets could be joined where it was difficult or impossible to access both sides of the sheets to be joined.

b   Of the six million parts that make up the Boeing 747, about three million are fastenings.

c   One of the most common fasteners used in the construction of aircraft is the rivet.

i   _____

_____

Each one of these screws, bolts, rivets, washers and clips has to be specified exactly, if an airworthy plane is to be produced. Not only must these fastenings be manufactured to internationally recognised standards, they must also be fitted in accordance with the maker's recommended procedures.

ii   _____

_____

Not only do rivets have nearly as good shear strength as bolts, but they are also usually lighter and easier to fit in large quantities. They also offer better resistance to vibration because they completely fill up the hole into which they are fitted. Stainless steel, aluminium alloy or titanium rivets are used, depending on which parts of the aircraft they are to be fitted in.

iii   _____

_____

However, it is now widely used in both structural and non-structural parts of aircraft in order to save man-hours. There are different types of blind rivet for different jobs, but the basic principle is the same.

Vocabulary

Find a word or phrase in the text above which means:

a   in a suitable condition to fly   _____

b   standards which apply in all countries   _____

c   ability to resist shearing strain   _____

d   ability to resist shaking   _____

e   aircraft parts that are not part of its structure   _____

⮕ **Workbook pages 25/26**

Review

Here is a picture of an aircraft being built from a kit.

Speaking and listening

1 **Discuss these questions in pairs.**

   a Why would someone want to build their own aeroplane?

   b Where would they build it?

   c How much do you think it would cost to buy the kit or the parts?

   d How many man-hours would it take?

   e What tools and facilities would they need?

> **Language Box**
>
> *would*
>
> To describe imaginary situations, use the form *would* + infinitive:
>
> *It would take a long time.*
>
> *I would like to build one.*
>
> *You would need a runway.*

2 🎧 **Listen to the interview with a man who built his own plane and make notes on the questions in Exercise 1.**

Vocabulary

1 Look at the following list of words from Unit 2. Put them into groups according to the number of syllables they have. For example, *engine* has two: en-gine; *micrometer* has four: mi-cro-me-ter.

technical   experience   interested   rivet   transmit
elements   grip   alloy   special   hammer   drill   lathe

1 syllable ●	2 syllables ● ●	3 syllables ● ● ●	4 syllables ● ● ● ●

2 Look at the Skills Box. Now underline the stressed syllable in each word in columns 2, 3 and 4 above.

> ## Skills Box
> **Word stress**
>
> It is extremely important in English that you stress the correct syllable; that is, you make one vowel louder, longer and higher than the others:
>
> ● the first syllable is stressed in en-gine;
> ● the second syllable is stressed in com-pu-ter;
> ● the third syllable is stressed in man-u-fac-ture.

Language

Three of the sentences below have grammatical mistakes in them. Correct the ones that are wrong.

a Tin snips are usually used for cut sheet metal.

b It's better to use blind riveting when you can't reaching both sides of the job.

c Because it is light, aluminium is especially useful in aircraft manufacture.

d Some people say that hand tools last a longer time than powered tools.

e To build it yourself, you would need good plans.

➪ Workbook pages 27/28

External structure

Speaking and vocabulary

Although these aircraft look very different from the outside, they have quite a lot in common.

1   **What are the similarities and differences between the aircraft in the pictures?**

2   **Look at the definitions below. Use the glossary to match them with the words in the box.**

> fuselage   wings   landing gear   rudder   struts   power plant   propeller   skin   fin   flaps

   a _____ large surfaces which project horizontally from the main body of the plane

   b _power plant_ the equipment which moves the aircraft forward

   c _____ equipment which supports the aircraft on the ground

   d _____ the main longitudinal body of the plane

   e _____ the covering surface of the main body

   f _____ strong rods or bars that are attached to the wings

   g _____ small, hinged control surfaces on the wings

   h _____ fixed vertical surface at the back

   i _____ pitched blades on the front of the power plant

   j _____ hinged vertical surface at the back of the plane

3   **Mark the stressed syllable in each word in the box in Exercise 2 and practise saying the words aloud.**

Language

1 Look at Language Box 1 and complete the table.

   a They are rods attached ____ the wings.

   b It's a hinged, vertical surface ____ the back of the plane.

   c It supports the aircraft ____ the ground.

   d These blades are ____ the front of the power plant.

   e The seats are ____ the cockpit.

2 Write a sentence of your own using each of the items from the table. Memorise the combinations with *at, the* and *on the*.

Speaking

1 Look at Language Box 2. Then expand the prompts below to make fuller descriptions.

   i These / large surfaces / project horizontally / main body / plane

   ii This / equipment / supports / aircraft / ground

   iii This / main longitudinal body / plane

   iv These / strong rods / bars / attached to the wings

   v These / small, hinged control surfaces / wings

   vi This / fixed vertical surface / back

   vii These / pitched blades / front / power plane

   viii This / hinged, vertical surface / back / plane

2 In pairs, one partner should define accurately a part of the aircraft structure; the other will say which it is.

Example:

   These / large, hinged control surfaces / wings

= These are large, hinged control surfaces *on the* wings.

⇨ **Workbook pages 29–31**

Lesson 2
Force

Speaking

How does each of these move? Discuss in pairs. Using words from the box, think about the directions that each one needs to travel in.

helicopter   humming bird   bee

> take-off climb dive bank hover
> roll fall lift forwards backwards
> rotate vertical horizontal

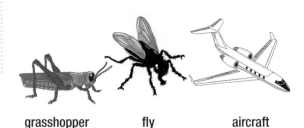

grasshopper   fly   aircraft

Reading

1   **Read the text below quickly. What types of movement does it talk about? For example, *lift*.**

2   **Find and circle the following words in the text as quickly as you can.**

> anticlockwise clockwise direction
> forces gravity lift roll or spin
> speed stresses the ground

Skills Box
Scanning

There is no need to read every word carefully in a large body of text. When people read in their first language, they often 'scan' the text, focusing on particular keywords, specific sentences and/or sections of the text. You need to train yourself to do this when you read in a different language.

Forces in motion

Everything that flies has to withstand a variety of different forces. The first, most obvious one is the force of gravity. To counteract gravity, any flying machine must have a way to provide lift. But that is just the beginning.

5   A machine or animal that only hovered above the ground would not be much use. Flying machines and animals have to go forwards, backwards, up, down and sideways when they are in the air. They also need to be able to accelerate, decelerate and roll or spin.

According to Newton's first law of motion, a force is required to make something change direction and speed. But according to his third law of motion, every time a force

10   is exerted there is an equal force in the opposite direction. For example, when the propeller of a plane is rotating clockwise, there is an equal and opposite force trying to turn the plane anticlockwise. It is this opposition of forces which produces the various stresses on the structure of aircraft, birds and insects alike.

3 What verbs in the text go with the words and expressions in Exercise 2? Find them in the text and write them next to the words. For example, *produces* goes with *stresses* in lines 12/13.

4 The stresses that aircraft have to withstand can be divided into five basic types, illustrated in the drawings *a* to *e*. Match the pictures with the descriptions below.

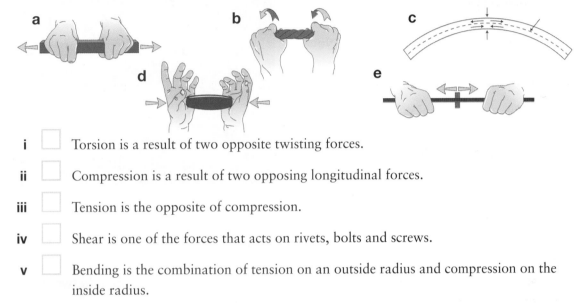

i ☐ Torsion is a result of two opposite twisting forces.

ii ☐ Compression is a result of two opposing longitudinal forces.

iii ☐ Tension is the opposite of compression.

iv ☐ Shear is one of the forces that acts on rivets, bolts and screws.

v ☐ Bending is the combination of tension on an outside radius and compression on the inside radius.

Vocabulary

1 Complete the sentences below, using words for the different forces. You may need to change the form of the word.

a Data _____ involves decreasing the volume of data so that it takes up less space on your computer.

b _____ tests are performed by twisting one end of a test specimen while the other is held fixed.

c You should _____ your knees when you lift something heavy.

d If there is too much _____, the rope will break.

e Bolts and rivets have good _____ strength.

2 Memorise some of the vocabulary from the lesson.

Speaking

1 Work in pairs. Think of two examples of each of the forces in *a* to *e* in Vocabulary above. One example should be from daily life and the other from engineering.

2 With a new partner, take turns to describe the example. Your new partner will say which kind of force is at work.

➪Workbook pages 31/32

Manoeuvres

Speaking

1 How many ways of moving can you remember from Lesson 2?

2 Look at the photo. What do you think are the special features of this bird? How does it need to fly?

Vocabulary and listening

Mr Lewis is visiting Kenya. He is talking to his guide in the early morning.

1 🎧 Cover the text of the dialogue opposite. Listen to the conversation between Mr Lewis and his guide and answer these general questions.

    a Where are they?

    b What are they looking at?

    c Who is the expert, Mr Lewis or the guide?

    d What is going to happen next?

2 🎧 Look at the words and phrases below. Listen to the second part of the conversation and number the phrases you hear in the correct order.
Note: You will not hear every phrase.

    a ☐ gain height

    b ☐ rotate

    c ☐ pull higher

    d ☐ take off

    e ☐ turn around

    f ☐ turn right

    g ☐ fall to the ground

    h ☐ climb away

    i ☐ dive straight down

3 Summarise the movements of the eagle.

4 🎧 **Look at the text of the first part of the conversation. In each line of the text, there is a word which is not in the conversation. Listen again and delete the extra word.**

> MR LEWIS: No, it's no good … I just can't see it. My eyes just aren't as good as yours. There's not really enough light for me yet.
>
> GUIDE: Don't worry. Be patient. Close your eyes and then look back across the lake again. He's on a branch about halfway up that very tall tree on the left side
> 5 of the rocks. Start at the bottom of the tree there and move your eyes slowly up until …
>
> MR LEWIS: Ah, yes, there he is! My gosh, look at him … if he only just stays still long enough while I get my binoculars. There, yes, I can see him clearly now … every feather. My, he's a big chap.
>
> 10 GUIDE: Yes, fully grown … and he's going to start his hunting in a minute, I think. Don't use your binoculars; you won't be able to follow him around. He's much too fast. Do you see those circles on the water surface over to your left? That's a very big fish just coming up to feed on the early morning insects. I'm sure the eagle must have seen him and he'll probably – yes,
> 15 indeed, there he goes.

5 🎧 **Listen once more to the second part of the conversation. When you hear the following words or phrases, say 'Stop!' Your teacher will stop the recording and replay only that sentence. You must write down exactly what is said in each case.**

a shallow angle   **d** special manoeuvre

b straight down   **e** fast

c heavier

6 Before your teacher gives you the correct answers, compare your sentences with a partner. Together, use your knowledge of the language to decide whether the spelling and grammar are correct.

Speaking

1 Mark the stress on the sentences in Exercise 5 above and practise saying them.

2 Look at the Skills Box. In pairs, look at the transcript and mark the sentence stress on a short section.

3 With your partner, practise that section, focusing on correct sentence stress.

➡ **Workbook pages 32–34**

Skills Box

Sentence stress

In a sentence in English, the words which carry the central message are stressed – they are spoken louder, longer and/or higher. Compare:
It's a _female_.
They are usually _bigger_ and _heavier._

Skeletons

Reading

Sea eagle
skeleton

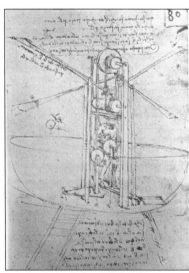

Leonardo da Vinci's
man-powered flying machine

Hawker Hunter
fighter

1 **Talk in pairs. How are the skeletons in the three pictures similar and how are they different in terms of:**

a function?

b construction?

2 **Look very quickly through the text below. Is it about one, two or all of the pictures?**

Skeletons in birds have a harder job than those of other animals. A bird's **internal** structure needs to be light enough for flight but also strong enough to bear the stresses of flying. To solve these problems, birds' skeletons have some special characteristics.

Birds' skeletons are extremely light in **relation** to their size. The frigate bird, for
5 example, which flies long distances over tropical oceans, has a **wingspan** of over two metres, but its skeleton is lighter than its feathers. Birds also have lightweight beaks instead of heavy teeth and jawbones. Some other bones are very small, or – like the tail bones – non-existent. The bones of a bird's main limbs are **hollow**, with special struts inside them to strengthen them. At the same time, other bones are even more
10 rigid then those of a mammal's skeleton. Stiff, horizontal **extensions** protruding from the bird's ribs lock them tightly together at the breastbone, sometimes called the wishbone (the next time you have roast chicken, notice its huge breastbone, which sticks out like the keel of boat). This bone is **unique** to birds. It anchors the huge **muscles,** which are the power plant that they need for flying.

Reading and writing

1 **Underline the most important information in the text.**

2 **Complete the outline summary of the text below.**

Birds' skeletons:
Need to be ______________ and ______________
Special characteristics:

____

____

____

Large breastbone which ______________________________

3 **Look at Skills Box 1. Give your writing to a partner to check the spelling and grammar.**

> **Skills Box 1**
>
> **Checking information**
>
> It is very important that you can see errors or items accidentally omitted from information, and you need to practise this skill.
>
> It will also help you to check work that you have written yourself.

Pronunciation

1 **Look at the highlighted words in the text.**

a How many syllables do they have?

b Which vowel sound does the stressed syllable have?

2 **Read Skills Box 2.**

a Check the pronunciation of the words in the box below.

b Add words to the correct column of the table below, according to the underlined vowel sound.

~~height~~ ~~around~~ straight flight
tail power dive away weight
~~gain~~ down climb ground

/aɪ/	/eɪ/	/aʊ/
height	*gain*	*around*

⇨ Workbook pages 34/35

> **Skills Box 2**
>
> **Pronunciation:** /aɪ/ /eɪ/ /aʊ/
>
> It is very important to pronounce and hear correctly combinations of two vowel sounds:
> /aɪ/ h**ei**ght d**i**ve cl**i**mb
> /eɪ/ t**a**ke str**ai**ght d**ay**
> /aʊ/ s**ou**nd d**ow**n am**ou**nt
>
> In these combinations (called diphthongs), the first vowel sound is stronger than the second, but both must be heard.

Lesson 5
Internal structure

Speaking

The fuselage of an aircraft, right, is a semi-monocoque design, which consists of a central framework, or skeleton, covered by stressed skin panels.

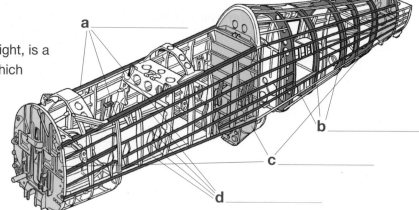

Work in pairs or groups.
What could be the advantages of:

a a purely monocoque design, in which the skin takes all the stress and there is little internal skeleton?

b a trussed airframe design, in which the frame carries all the stress and the skin none at all, as in the wood-and-fabric aircraft of the pre-1930s?

Reading

1 In the following text about the fuselage of the aircraft, quickly underline all the occurrences of these words: *aircraft*, *skin*, *longerons*, *frames*, *bulkhead*, *fuselage*.

2 Read the text slowly and label the parts A to D on the drawing. Some are the same.

The skin panels are an integral part of the structural strength of the aircraft and are attached to the three main components of the internal framework, which are:

- **Frames:** These are the eleven lateral members that are located at intervals along the length of the fuselage and maintain the shape of the aircraft. Some frames are in
5 the form of closed rings which run right round the inner surface of the skin. Some frames are half-ring or U-shaped to allow access from the top of the fuselage. Others are in the form of bulkheads, which act as both protective shields and mounting plates.

- **Longerons:** There are four of these, extending lengthwise from the back of the
10 cockpit to the stainless steel fireproof bulkhead behind the engine. The longerons provide longitudinal stiffening in the forward fuselage. They also transfer forward thrust from the engine to the fuselage structure.

- **Stringers:** These bars, thinner and lighter than the longerons, run longitudinally between the frames to provide additional stiffening.

Vocabulary

Look at the drawing and description of the wing.

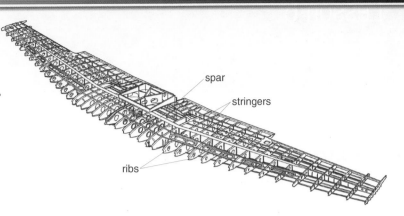

This wing is a cantilever design without any bracing from external struts. The main structural members of the wing are:

- **Spars:** These run **laterally through the wing** and **act as the main support** for the wing structure.
- **Stringers:** These thinner bars **run** parallel to the spars.
- **Ribs:** These are longitudinal plates connecting the spars and stringers.

Like the fuselage, the structure is completed by the addition of a stressed skin, attached to all the internal members.

1 The four words and phrases in bold above describe the structure of the airframe. Put each of them into one of the columns below.

airframe part	verb	direction	purpose

2 Now look back through the description of the fuselage in Reading Exercise 2 and add four more words or phrases to each column of the table.

Writing and speaking

1 Work in pairs, A and B. Student A should look at the information on page 244. Student B should look at the information on page 245. Make notes on how this type of aircraft is constructed.

2 Give a short talk on your topic. You can use your notes but you may not read from them. Your colleagues must ask questions at the end.

➡ Workbook pages 36/37

Lesson 6
Assembly

Speaking and listening

You are going to hear a talk about an aircraft production line.

1 Before you listen, look at the picture below of a production line. Work in small groups. Discuss the stages of production you would expect to find at a factory like this. In what order would these stages happen?

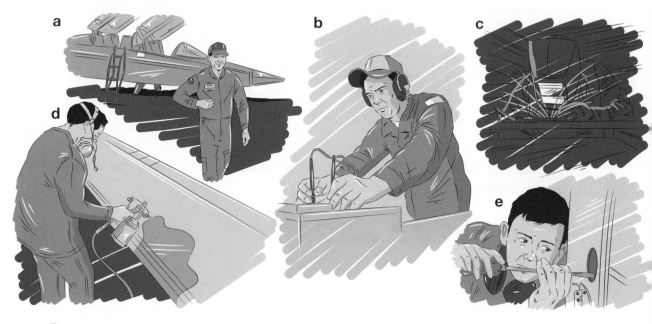

2 🎧 Listen to the description of the production of a jet fighter. While you listen, number the stages in the correct order.

a ☐ engines

b ☐ main assemblies

c ☐ painting

d ☐ flight control systems

e ☐ hydraulics tests

f ☐ pre-flight tests

g ☐ wings

h ☐ flight tests

i ☐ electronics tests

Skills Box 1

Predicting

You can always help your understanding of spoken English by using visual clues and your knowledge of the world.

3 Talk to a partner. Was the order the same as you expected? What was similar/different?

Vocabulary

🎧**Look at Skills Box 2. Then listen again and correct the sequencing signposts which the speaker uses.**

a In the first stage   _____

b Nextly   _____

c At these points   _____

d At third stage   _____

e Following on satisfactory
   completion   _____

f Following to this   _____

g Once this has been carried out _____

Language

1 **Use the passive form of the verbs below to complete these sentences.**

> test splice paint install attach carry out

a The three main assemblies _____ together.

b The aircraft _____ in air force colours.

c Pre-flight tests _____ .

d The wings _____ to the fuselage.

e Navigation systems _____ .

f Electrical cables _____ .

2 🎧 **Listen to the talk again and check your answers.**

3 **Look at the picture of an aircraft during construction. Ask and answer questions about it using the passive voice. For example: *Has the cockpit been fitted?* Use the prompts.**

cockpit fitted?

main assemblies spliced together?

flight control surfaces attached?

engines assembled?

weapons systems fitted?

wings painted?

pre-flight tests carried out?

moved to Station 4?

➡**Workbook pages 37/38**

Lesson 7
Production lines

Speaking

In the past, aircraft were usually built *in situ*. However, although this still happens with small light aircraft, most commercial and military aircraft are now manufactured on a production line which is constantly in motion.

1 Look at the photograph on the opposite page. In pairs, discuss what processes you can see.

2 Discuss the following questions with a partner. Use some of the speculating words and phrases from the Skills Box.

    a How many parts are there in a large airliner?

    b How are the parts transported to and around the factory?

    c How long does it take to build an airliner?

Skills Box

Speculating

When you are guessing or unsure, one option is to use *maybe*, for example: *Maybe they use robots.*

There are many other common alternatives which you can use to express your ideas more exactly, for example:
*They **must** have about 150.*
*It **probably** costs a lot.*
I'd say it'll take about two weeks.
*Yes, **perhaps**.*
I doubt if they do it by hand.

Reading

Look at the text opposite.

    a In no more than 30 seconds, circle all the numbers in the text.

    b Some of the numbers in the table are incorrect. Check and correct the information.

    c Were your answers in Speaking Exercise 2 right?

a	number of homes that could fit inside the factory	200
b	hangar door width	900 m
c	parts in a 777 airliner	300,000
d	bridge cranes	25+
e	build time for a 747	±150 days
f	shifts	3
g	employees	2,000
h	maximum crane load	40 tons
i	height of hangar doors	90 ft

At the Boeing aircraft production site at Everett in the USA, every 747 airliner starts life as wing spars at one end of the
5 production line and ends up at the other as a completed aircraft, on average four months later.

The main assembly building at Everett is the largest building in
10 the world by volume. It is so big that about 2,000 family homes could fit inside. In addition to the 747, the 767 and 777 aircraft are produced here. At least 25,000 people are employed here working three shifts, and the site is so large and complex that it has its own fire department, fully equipped health clinic and
15 electrical substations.

The 767 and 777 each consist of around three million parts, and the 747 has twice as many as this. Ordering, tracking and distributing the thousands of components and sub-assemblies are extremely complex management tasks. It is impractical and uneconomic to keep components in stock all the time, so a system of last-minute
20 delivery is used. This method, called 'just-in-time', ensures that parts only arrive at the factory just before they are needed and can be delivered to the right point on the assembly line at exactly the right time.

Every day, parts and sub-assemblies are received from all over the world as well as other parts of the USA. Many of them arrive by ship, and there is a railway line that
25 runs directly from the nearby port into the factory. Once inside the factory, the parts are distributed by fork-lift trucks and cranes. There are more than 25 overhead bridge cranes, some capable of lifting up to 40 tons. These move backwards and forwards some 90 feet above the floor, delivering heavy items from one assembly position to another.

30 As production proceeds, the aircraft move closer and closer to the hangar doors at the end of the line. Finally, these enormous doors, over 26 metres high and 90 metres wide, are opened so that the finished plane can be rolled out onto the tarmac, ready for flight testing.

Speaking

What is the most surprising/interesting thing about the Boeing plant? Discuss your ideas in small groups.

⇨ Workbook pages 39/40

Lesson 8
Robots

Speaking

1   Which of the following do you think will happen in your lifetime? Mark each one with a number between 1 and 5, where 1 = definitely and 5 = definitely not.

a   ☐   Robots will do all our work for us.

b   ☐   There will be robots that can detect human emotion.

c   ☐   Robots will be used in the home.

d   ☐   Robots will be used to do more difficult and sophisticated jobs.

e   ☐   People will be replaced by robots in the manufacturing industries.

f   ☐   More money will be invested in robot design.

2   In pairs, discuss your answers and explain your reasons.

Reading

1   Read the text very quickly and answer the questions.

a   What are snake-arm robots?

b   Will you change any of your answers to Speaking Exercise 1 after reading the text?

> **Snake-arm robots to investigate a new approach to aircraft assembly**
>
> OC Robotics has secured a contract from Airbus to develop snake-arm robot technology for possible aircraft manufacturing processes. The focus will be assembly tasks within wing boxes – an area currently inaccessible to automation.
>
> 5   The composite, single-skin construction of aircraft structures, such as the A350 wing, presents new challenges for robotic assembly. Today, an aircraft fitter climbs into the wing box through a small access panel and uses manual or power tools to perform a variety of tasks.
>
> OC Robotics' snake-arm robots provide the opportunity to replace manual procedures
> 10   by delivering the required tools to all areas of the wing box. They could be used to perform tasks such as final sealant application and swaging.
>
> More widely, the development of snake-arm robots could ultimately enable major design and process changes, creating the opportunity for considerable cost savings for the aerospace industry. Future wings could be designed with fewer and smaller access
> 15   panels and maintenance times could be reduced.
>
> On successful completion of trials, the industrialisation process will be completed in the following 12 months.   *Reproduced with permission from OC Robotics. www.ocrobotics.com*

2 Look at the Skills Box. Which elements helped you understand the main idea of the text in Exercise 1?

Speaking

1 Work in small groups. Your team has been asked for its opinion on the best area of research into robots.

industry: e.g., construction, design, maintenance

the home: e.g., security, maintenance, pets

business: e.g., working with customers

a Choose the area where you think robots could be the most useful. Be clear about the reasons for your choice.

b Decide, in as much detail as possible, what the specific challenges will be for artificial intelligence in that area.

2 Form new groups with members of the other teams. Explain your proposal to the members of your new group.

3 In your new groups, decide which area would be best to research. Present your decisions and your reasons to the class.

Language

1 Can you say the following sentences in such a way that they have the meaning in brackets? (You shouldn't use the word in the brackets.)

a the road traffic problem/worse/before/better (definitely)

b if world climate/warmer/serious problems (perhaps)

c in future/everybody/live for 120 years (I'm not so sure)

d space tourism/popular/during this century (probably)

e poverty/eradicated in the 21st century (definitely not)

f the population of the world/pass ten billion (probably not)

g in my work/have to fly (I don't know)

2 Do you agree with the statements? Discuss in pairs.

➪ Workbook pages 40/41

Skills Box

Skimming

You need to practise the skill of looking quickly through a large amount of text to get the general idea of it. Elements in the text which can help you are:

titles
layout
capitalised words
numbers
key nouns
pictures

Language Box

Discussing probability

- *Will* is used in English to express strong commitment to a probability. It is often used to predict future events, for example:
  If there are no more delays, we will finish in October.
  We *will not/won't* get there on time.
- It's also used to talk about future facts, for example:
  The tests will begin next week.
- *May, might* and *could* are used to talk about future probability:
  Robots may replace people in many everyday jobs.

Roles on the assembly line

Speaking and reading

1   Look at the jobs below quickly and find out what duties and skills each one requires.

VACANCIES

a

Position: Aircraft Technician
–2-year renewable contract
Location: Oman
Salary: from £21,500
Important: Must have structural/sheet metal background
Job specification:
We are looking for an aircraft technician to work on predominantly Hawk aircraft in Masirah, Oman. The successful candidate will need to have experience of all aspects of aircraft structures. You will be required to complete modifications on the aircraft and fabricate parts, etc., working from drawings and manuals. You must have riveting experience. Previous experience on Hawk is not essential. Candidates with commercial experience will be considered for this position.

£17,033 basic + £4,000 overseas allowance + £500 location bonus ALL TAX FREE. Medical and dental cover, free accommodation in the Sergeants' mess.

b

Position: Mechanical Design Draughtsman
Location: Lancashire, UK
Rate: £23.75/hour
Duration: 6 months
Job specification:
To produce manufacturing and design data to enable manufacture and assembly of clients' products.
Experience:
Some or all of the following:
Sheet metal weatherproof and non-weatherproof enclosures, including the use of inserts. Structural steel detail design; awareness of prevention of galvanic corrosion.
Complex CNC-machined components. A thorough understanding of the need to record modifications to drawings and documents.
Desirable:
HNC in mechanical engineering, competent EUCLID & AUTOCAD user.

c

Wanted: Stress Engineers
– permanent
Location: Kuwait
Salary: £20,000 to £40,000
Job specification:
Important: Degree with a minimum of 1 year's industry experience of stress engineering.
Some excellent opportunities exist for stress engineers, requirements vary from novice engineers to experienced stress engineers to work with a prestigious company based in Kuwait. These opportunities offer an exciting and dynamic work environment. The main tasks will be to perform stress analysis in the form of finite element (FE) or fatigue and damage tolerance and ideally you will have previous experience of working on airframe platforms. You will be required to work closely with the customer and so must be a good team player & able to work in a pressurised environment with minimal assistance. You will be required to deliver to time, cost and quality. You must be PC-literate and be able to use Mathcad and Microsoft Office (Word, Excel).

d

Position: Systems Engineer, Navigation
– permanent
Salary: £35,000 to £42,000
Important: Algorithm and mathematical bias
Location: South Wales
Job specification:
You will be working on a jet trainer project. Aircraft is single-engine, smaller than a Hawk. The candidate will be the single point of contact for any system queries on this project. Project is at test stage, although candidates will need to have some requirements experience as the product will be handed over to customer for review and changes are likely. Candidates will be reviewing flight tests, steering tests, systems test and mission systems. Candidates must be able to liaise with other departments and the customer. A mathematical, navigation, steering and weapons-aiming background would be useful. Candidates will have plenty of opportunity to progress.

2   Read the advertisements more closely. Which job would you prefer? What would be the disadvantages of that job for you?

3   Discuss your choice in pairs.

Listening

These four people are looking for a job in the aircraft industry.

Qualifications	Qualifications	Qualifications	Qualifications
Background	Background	Background	Background
Experience	Experience	Experience	Experience
Reasons	Reasons	Reasons	Reasons

a ☐   b ☐   c ☐   d ☐

1   🎧 Listen and choose a suitable job for each person.

2   🎧 Listen again and take notes for each candidate in the space under his picture.

3   Compare notes with a partner.

Speaking

Work with a new partner. You are going to interview him for one of the jobs, *a* to *d*.

a   Find out which job he would like to apply for.

b   Prepare ten questions to ask him.

c   Interview your partner.

➡ Workbook pages 42/43

Skills Box

Listening to a talk

Notes

When you are taking notes from a talk, remember: don't try to write down everything; use abbreviations and numbers.

Accents

In your work, you will often listen to people from all over the world speaking English. It is important that you get used to this and make sure your own accent is easy to understand.

Speaking clearly

Working in aircraft production or maintenance invariably involves working with people of different nationalities, roles, skills, experience and standards of education. Consequently, it is extremely important to try to communicate properly. A company with effective systems of communication is a safer, more efficient and happier place to work.

Vocabulary

1   Look at the words below and decide what parts of speech they are: noun (*n*), verb (*v*) or both (*b*).

   a log ___   **c** intercom ___   **e** leaflet ___   **g** instructions ___

   b terminology ___ **d** repeat ___   **f** order ___   **h** shout ___

2   Now write the words in the correct place in the table.

connected with written communication	connected with spoken communication	connected with written and spoken communication

Reading and speaking

1   Divide into five groups, *a* to *e*, and read the corresponding paragraph carefully.

2   Prepare a short talk about your aspect of communication, including the information in the text and any other information you want to add.

3   Form groups of five with one student from each group, *a* to *e*. Give your talk. Your colleagues will listen, take notes and ask questions at the end. These notes will be important later.

a　In work situations, you should always try to use the correct terminology. Unless you use the special words that are related to the job, you may well cause confusion. If someone uses terminology that you don't understand, you should always ask them to explain what they mean. It is very important that you should never be afraid to tell
5　someone you don't understand. A few minutes which you spend getting something clear in your mind can save many hours of work in the long run.

b　Sometimes, you may have to give orders or instructions. These should be as short and simple as possible. Sometimes, it is a good idea to check that you have been understood by asking for your orders to be repeated. If you are given an order, it is
10　often a good idea to repeat it so that the person who gave it can spot any mistakes. It is also vitally important not to pretend to understand what you are supposed to do just to save face. If you are not sure, ask for clarification. When unusual or emergency situations arise, clear communication is especially important: all relevant information must be reported to the supervisor and other employees so that they can act quickly
15　and correctly.

c　You should be especially careful when using the telephone, intercom or radio. It is usually a good idea to repeat information, whether you are giving it or receiving it, and to make a written note of it. Many people shout on the phone because they think others will hear them better. Try not to do this: shouting causes distortion. Care must
20　be taken with mobile technology, however, that it does not interfere with machinery.

d　The log is an important link in the chain of communication. It is a way of passing information from one shift to the next to ensure safety and smooth operations. If you have to complete a log, you should make clear and accurate entries and be sure to write down anything unusual that has happened. When changing shifts, it is a good
25　idea to have a short talk with your relief. He should be able to read the log and ask questions before you go, to help him to understand any problems that may arise.

e　One way that manufacturers are improving communication in the workplace is by dividing the workforce into maintenance or production cells. At one factory where the Eurofighter Typhoon is produced, this concept of *cellular manufacturing* has been
30　introduced in order to give quality the highest priority. Each cell is almost completely self-managed and is responsible and accountable for its own performance. Initial communication about technical and personnel problems takes place within the team, which usually means that difficulties can be overcome more quickly. The aim is to create a feeling of responsibility and pride in the cell's product.

Writing

Prepare a 'Good Communications' information leaflet or notice for the manufacturing workplace, using your notes but without using the text for help.

⇨Workbook pages 44–46

UNIT 4

Lesson 1
The power of water

Reading and speaking

1 **In pairs, think about the places where you live and work and discuss these questions.**

  a What is water used for? For example, *bathing*.

  b How is it supplied to where it is needed?

  c Are there any geographical and technical difficulties in supplying water to some places?

2 **Scan the following text. Answer the questions from Exercise 1 for the Alhambra Palace in Granada.**

Anyone who has visited the beautiful Moorish palace of the Alhambra in Granada will know that as well as the complex and beautiful design of the buildings, there are also fountains and pools which provide relief in the dry 40°C
5 summer heat. The Arab engineers and architects who built it were aware of the importance of water, not only for drinking and bathing, but as a source of beauty and power.

When they were originally built, these fountains were powered by natural water pressure. The supply of water came from the Darro River, which has its source high
10 up in the mountains of the Sierra Nevada. The water was raised from the river and stored in reservoirs which had sufficient capacity to provide a constant supply even when the river level was very low. The potential energy of the water was converted to kinetic energy by a complex system of channels and pipes. The flow rate was determined by the dimensions of the channels and pipes and the gradient of the slope
15 down which they ran.

The gardens of the Alhambra – in Spanish, *el Generalife* – are just one example from the 'golden age' of technology enjoyed by Islamic society between the ninth and 16th centuries. The development of cities such as Granada, Cordoba, Baghdad and Marrakesh depended on increasingly sophisticated methods of water management not
20 only to ensure a clean supply for drinking and washing, but also to drive machinery and to provide farms with water.

Among the many Islamic thinkers who contributed to hydraulic technology, probably the most famous is the 13th-century engineer Al-Jazari. As well as clocks and water mills, he designed water pumps that used the power of naturally flowing water to
25 drive reciprocating pistons which sucked water into a cylinder through inlet flap valves and then pushed it on upwards through outlet flap valves. The outlet valves closed on the suction stroke and the inlet valves closed on the pumping stroke.

3 With a partner, discuss the similarities and differences in meaning between each of the following pairs of words from the text.

    a pool/reservoir

    b channel/pipe

    c capacity/dimensions

    d level/gradient

    e sophisticated/complex

    f suction/pumping

Vocabulary

1 Match each word on the left with one from the right to make common two-word combinations. All the combinations are from the reading text.

a	water	**i**	piston
b	constant	**ii**	pressure
c	kinetic	**iii**	rate
d	water	**iv**	supply
e	flow	**v**	energy
f	reciprocating	**vi**	valve
g	inlet/outlet	**vii**	management

2 Use the two-word combinations to complete the following sentences.

    a _____ is normally measured in psi (pounds per square inch).

    b The _____ of a liquid is determined by the dimensions of the channels and pipes it runs through.

    c _____ are used to control the flow of water in and out of pumps and tanks.

    d Reservoirs can ensure a _____ of water even when the water level is very low.

    e A _____ sucks water into the cylinder of the water pump.

    f _____ is the energy that something possesses because of its motion.

    g Some ancient civilisations used sophisticated systems of _____.

3 Look at the Skills Box. Check that you can pronounce all the phrases in Vocabulary Exercise 1 correctly.

➯ Workbook pages 47/48

Skills Box

Two-word stress

In two-word combinations, each word usually keeps the same stressed syllable which it always has.

However, if both words in the compound are nouns, the first word normally takes more stress than the second, e.g.:

a <u>project</u> meeting

<u>water</u> supply

an <u>aircraft</u> factory

Speaking

1 The pictures below show two different types of ram. Are there any similarities between them?

2 Discuss whether these sentences about hydraulic ram pumps are true or false.

- Hydraulic ram pumps work on the principle that liquid cannot be compressed.
- They always involve the movement of water in a closed system.
- When force is applied to one point of the system, it is transferred to another point by the fluid in the system.

Vocabulary and listening

1 Check that you remember the meaning of these words. Which ones are represented by the symbols?

inlet/outlet   spring   valve   increase   upwards   momentum   cycle

The diagram below shows the principal components of a hydraulic ram pump. This is a device which, like Al-Jazari's pump, uses water power to lift water above its natural level.

2 Study the diagram. Discuss with a partner how you think the diagram should be labelled.

a delivery pipe

b inlet pipe

c spring-loaded clack valve (non-return valve)

d delivery check valve

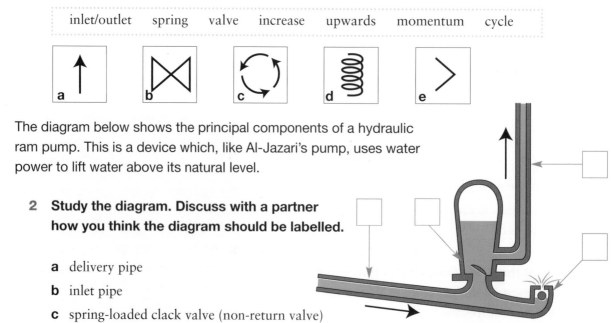

3 🎧 Now listen to a teacher explaining how the hydraulic ram pump works and check whether you were right.

4 🎧 Listen to part of the talk again and put the word at the end of the line in the correct place.

So when this happens to water, there is this backward pressure	the
in the pump, and this enough to open the delivery valve here in	is
the centre of the diagram push water upwards through the outlet	and
pipe. Of course, then the pressure inside the pump goes down,	again
5 and as it returns to normal, spring-loaded valve at the bottom – it's	our
sometimes called a 'clack' valve it makes a clacking sound like	because
someone hitting pieces of metal together ... actually, sometimes	two
it doesn't have a spring, just a weight – this valve opens again,	anyway
and of course the – what happens the delivery valve up here?	to

Speaking

1 Look at the Skills Box. Then mark the stressed words and all the weak /ə/ sounds in these phrases.

   a and the whole cycle starts again

   b it flows down a pipe on a slope

   c Yes, and when the water hits the ram

   d and it doesn't need an electric motor or an engine

   e the energy is changed into pressure

Skills Box

Weak /ə/

Where there is no stress on a syllable, the vowel sound is often the weak /ə/. The words *and*, *a*, *an* and *the* are not usually stressed and so are pronounced with a weak /ə/ sound.

the is pronounced /ðiː/ before a vowel sound.

2 Practise saying the sentences in Exercise 1.

3 Look at the text in Vocabulary and listening Exercise 4 above. Mark the stressed words and then, with a partner, practise saying the text aloud.

Writing and speaking

1 With your partner, prepare brief notes on how a hydraulic ram pump works. Use some of the words in the box.

> flow   gradient   ram   open/close valve   high/low pressure   opposite direction

2 Work with a new partner. Describe to him the principle of the hydraulic ram pump and make one or two deliberate, small errors. Your partner will listen and, at the end, tell you which parts were not correct. You may use your notes, but you are not allowed to read directly from them.

⇨ Workbook pages 49/50

Lesson 3
Modern hydraulics

There are two key facts that make hydraulic power important.

- Firstly, hydraulic devices can give a 'mechanical advantage'; that is, a large amount of work can be produced from a small amount of effort.

- Secondly, hydraulic systems can be used to transmit a force immediately from one place to another, without the force being reduced in strength.

Speaking

Look at the two diagrams. With a partner, discuss the mechanical advantage in each case.

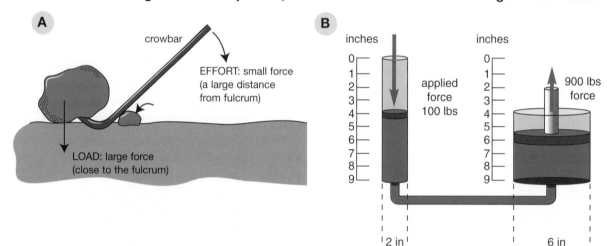

Reading and vocabulary

1   **Read the text opposite quickly and choose a suitable title for each paragraph.**

    a Hydraulics in the aircraft industry

    b Future hydraulic systems

    c The historical development of hydraulics

    d Everyday applications of hydraulics

    e Fluid properties

2   **Complete the gaps in the text using the words in the box.**

> technological   increasing   atmospheric   properties   fluid
> electrical   surfaces   loading   conditions   ram

The hydraulic (a)_____ pump is a very simple way of using the pressure of a liquid or fluid to do work, and there is hardly any aspect of modern (b)_____ life that doesn't involve hydraulics in one form or another. The effectiveness of a hydraulic system can be increased considerably by (c)_____ the pressure inside the

5　system, using a pump driven by the power plant of the equipment. The power-assisted steering found in most modern cars is an example of this, whereas the brakes used to stop the car work effectively with fluid at (d)_____ pressure.

Because of its unsuitable chemical and physical (e)_____, water is not normally used in modern hydraulic systems. It has been replaced by a number of

10　different kinds of hydraulic fluid, which are usually some kind of oil. The properties of the fluid will depend on the ambient operating (f)_____ of the system. A hydraulic system which operates in very low temperatures obviously requires a (g)_____ with a very low freezing point. Conversely, those operating in high temperatures must have a high boiling point.

15　Hydraulic systems are used throughout the aircraft industry. On the tarmac and in the hangar, much of the equipment used for towing, lifting and (h)_____ aircraft is hydraulic. Aboard aircraft, hydraulic systems are used to control the flight surfaces, landing gear and other ancillary equipment. Most aircraft have a combination of hydraulic, (i)_____ and mechanical controls, although some small light

20　planes are only fitted with mechanical linkages. For example, they may be fitted with mechanical linkages for the primary control (j)_____, i.e., the ailerons, rudder and elevators, but the flaps, air brake and landing gear may be hydraulically operated.

Language

1　Look at the Language Box. Underline the adverb in each of these phrases from the text.

a　there is hardly any aspect of modern life

b　the effectiveness of the system can be increased considerably

c　the brakes work effectively with fluid at atmospheric pressure

d　water is not normally used

e　which is usually some kind of oil

f　the flaps, air brake and landing gear are hydraulically operated

g　some planes are only fitted with mechanical linkages

> **Language Box**
>
> **Adverbs**
>
> Information about a verb must be added using an adverb. Adverbs give information about how, when or where the action happens, e.g.:
> *It needs to be done **accurately**.*
> *This was **originally** built in 1952.*
> *They are **often** wrong.*
>
> Adverbs often end in ~*ly*, but many do not, e.g.:
> *hard, now, well, sometimes, very*

2　Circle the verb which each adverb refers to.

3　Go back to the text and underline all seven of the phrases. Can you find any additional adverbs in the text?

⇨ Workbook pages 50–52

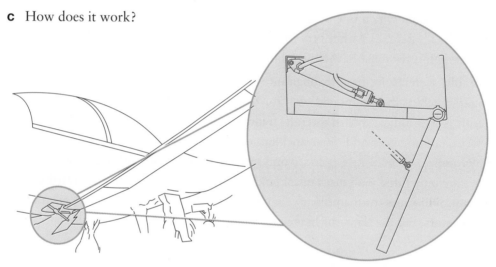

Lesson 4
The air brake

Speaking and reading

1 Work with a partner. Look at the drawing below and discuss the following.

   a What does it show?

   b What is the function of this system?

   c How does it work?

2 Now read the text in Reading Exercise 1 opposite and check whether you were right.

Vocabulary

1 Check the meaning of these verbs in your dictionary and compare your ideas with a partner.

> extend   retract   actuate   control   select   indicate

2 Complete the sentences about the air brake system by choosing the correct word from each pair.

   a The main control surfaces of an aircraft are operated by *actuators / indicators*.

   b The air brake system is hydraulically *operated / selected*.

   c It *actuates / extends* and retracts the air brake.

   d A hydraulic power system supplies hydraulic *fluid / liquid* to the system.

   e There are *operator / selector* switches and a valve in the cockpit.

   f These switches energise the selector valve *solenoid / piston*.

Reading

1   **Read the handbook description below and underline any words or phrases from Vocabulary Exercise 2.**

The air brake system is a hydraulically operated, electrically controlled system for extending and retracting the air brake. Hydraulic fluid at 3,000 psi pressure is supplied to the system by the hydraulic power system.

The air brake system comprises a selector and its associated switches, a selector valve,
5   a hydraulic actuator, a relay, an air brake indicator and its associated microswitch.

First, the selector valve solenoid is energised by the selector switch and causes the selector valve to direct the high-pressure fluid to the piston head chamber. This drives the piston, which is linked to the air brake, out of the chamber, thus forcing the air brake to extend below the fuselage. The microswitch is released and a light goes on in
10   the cockpit to indicate that the air brake is in the extended position.

The air brake is retracted by choosing the appropriate position on the selector switch. The valve solenoid is de-energised and the high-pressure fluid is redirected to the retract chamber. The system returns to the condition shown in the diagram, with the air brake fully retracted.

2   **Study the schematic below of the air brake system from a light aircraft and read the information again carefully. There are six spelling mistakes in the labelling of the schematic. Find and correct them.**

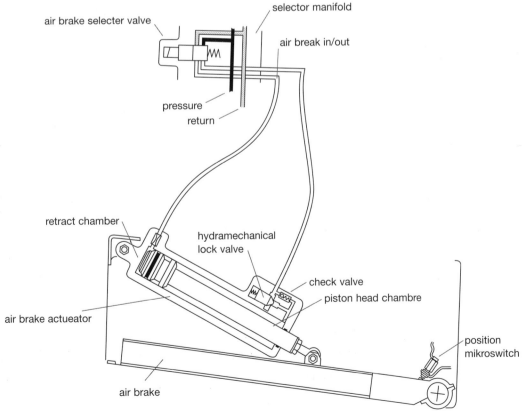

➡ **Workbook page 53**

Happy landing

Speaking

1 Cover the text on the right. Work with a partner and describe the stages in the animal's fall.

a

b

c

d

e

   i The head and front part of the body are twisted to face the ground, and the back legs are bent.

   ii It lands on all four feet to spread the load, and with the backbone slightly curved to absorb the shock.

   iii The front legs are brought up towards the face to protect the head from impact.

   iv It twists the back part of the body to line up with the front and extends all its legs.

   v The cat determines which way up it is before rotating its head so that it is the right way up.

2 Match the pictures with the five stages on the right.

3 Discuss this question with a partner: *Which aspects of the cat's solution to landing are applied to an aircraft?*

Vocabulary

1 Which words on the left have a similar meaning to those on the right?

a descent	**i** distribute
b determine	**ii** shock
c flexibility	**iii** fall
d impact	**iv** exact
e manoeuvre	**v** decide/judge
f precise	**vi** turn
g spread	**vii** mobility

2 Check that you can pronounce all these words correctly.

Listening

You are going to listen to a lecturer comparing aircraft landing systems with those of a cat.

1 🎧 **Listen first for the main ideas. In which order are the three things in the pictures discussed?**

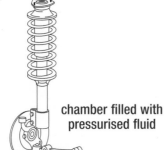

chamber filled with pressurised fluid

skeleton of a cat

inner ear of a cat

2 🎧 **With a partner, decide whether you think the statements are true (T) or false (F). If you're not sure, mark it with a question mark (?). Then listen to check answers.**

a ____ A pilot can determine his position relative to the ground quickly.

b ____ A falling cat slows down as it falls.

c ____ An artificial horizon is something in the inner ear.

d ____ Both a pilot and a cat use visual information to judge their landing.

e ____ Both always land on all points simultaneously.

f ____ Both have flexibility built into their internal structure.

g ____ Hydraulic systems do the same job in aircraft as muscles do in animals.

h ____ The wheels are filled with high-pressure liquid.

i ____ The landing gear acts as a piston.

j ____ The pilot can manoeuvre more easily at low speeds.

k ____ A cat's flexible skeleton helps it survive an emergency landing.

3 **Look at the Skills Box.**

a Look at the sketches on page 246 showing the process of an aircraft landing.

b Label sketches 3 to 7 in note form.

c Practise explaining your sketches to a partner. Remember, when you are speaking you must use full grammatical sentences, with the most important words stressed.

➡️ Workbook pages 54/55

Skills Box

Short forms for notes

When you are writing in note form, it is very common to leave out certain parts of the sentence: *a/an*, *the*, the verb *be* and the pronoun *I*.

For example:
The report is not finished yet.
Report not finished yet.

I will send it this afternoon.
Will send this afternoon.

This is also usual in passive sentences:
The legs are brought up to protect the face.
Legs brought up to protect face.

Note: When you are speaking, or writing longer texts, it is not normal to do this.

Douglas DC-3

Douglas DC-4

Most medium-sized and large aircraft today are built with tricycle undercarriages. However, there are some smaller aircraft that are fitted with the more old-fashioned taildragger design.

Speaking

Work with a partner and discuss these questions.

a What are the main functions of an aircraft undercarriage?

b Which of the aircraft in the pictures has a tricycle undercarriage and which is a 'taildragger'?

c Can you think of advantages and disadvantages of each of these designs?

Vocabulary

1 **Check that you know the correct pronunciation, including the stress, of these words and expressions. Use them to label the diagrams.**

centre of gravity   landing gear   cockpit   propeller   nose wheel

A

B

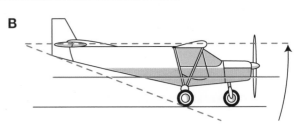

2 To which landing gear in Exercise 1 does each of these statements apply?

		A	B
a	Forward visibility is not so good.		
b	The centre of gravity is behind the front landing gear.		
c	It's easier to load and unload the aircraft.		
d	You don't need so much skill to land and take off.		
e	The nose wheel is easily damaged on rough and uneven ground.		
f	The plane can spin round (or groundloop) when landing or taking off.		
g	Because it is closer, the propeller sometimes hits the ground.		
h	The landing gear is a lighter load for the plane to carry.		
i	It can be damaged by very strong winds when parked on the ground.		
j	There is more air resistance, which means lower speeds.		

Listening

1 🎧 Listen to a discussion between two members of a flying club and check your answers.

2 🎧 Listen again to part of the conversation. Which part of the information below is incorrect in each case?

a There's no danger of breaking your nose wheel in a hole.

b I'd much rather make an emergency landing in yours than mine.

c The centre of gravity is between the wheels.

d And the cockpit is pointing up towards the sky – you can't see a thing.

e You just have to pay close attention.

f I have done a lot of groundloops myself.

3 Work with a partner and decide what design features in an aircraft would be best for each of the following situations. Think about: size of aircraft, number of seats, which landing gear design (tricycle or taildragger).

a For a small postal service between villages in the interior of the country.

b For an ex-air force pilot who likes traditional flying to use in his free time.

c For a young man who enjoys danger to fly as a member of a club.

⇨ Workbook pages 56/57

Speaking

1 Work with a partner. From Unit 3, can you remember the main control surfaces of an aircraft?

2 With your partner, discuss the meaning of the following notes, written by a student, about an aircraft's control surfaces.

> take-off – flaps part extended: lift without slowing
>
> landing – flaps fully extended: lift + air resistance (drag) = reduce
>   speed, but not height
>   Pilot decrease forward speed + control angle of descent

Vocabulary

Look at Diagram a.

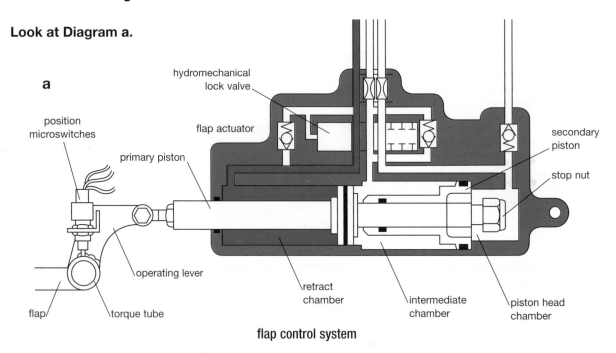

flap control system

1 **Divide the labels in Diagram a into three groups as follows: compound nouns where:**
   a both/all words are familiar to you.
   b one word is familiar to you.
   c both/all words are unfamiliar to you.

2 **Compare your groups with a partner's.**

Reading

Look at Diagrams b and c. How are they different from Diagram a?

b

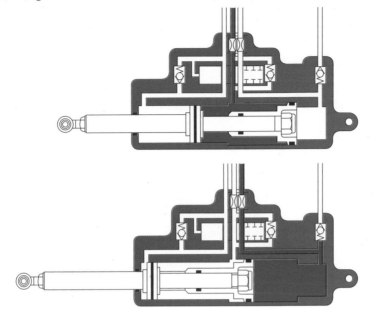

c

During take-off:

- The pilot selects the **TO** position.
- Hydraulic fluid at 3,000 psi is directed to the intermediate chamber.
- The primary piston is pushed out of the retract chamber so that the flaps move to
5 the take-off position. The pressure in the intermediate chamber also prevents any
movement of the secondary piston.

During landing:

- The pilot selects the **LAND** position.
- Hydraulic fluid is directed to the piston head chamber
10 • This pressure drives both the primary and secondary pistons into the **LAND**
position.
- The flaps extend to their maximum position.

After landing:

- The pilot selects the **UP** or retract position.
15 • Pressurised fluid is redirected from the piston head and intermediate chambers into
the retract chamber.
- This forces the pistons back into the retract position and the flaps return to their
neutral position.

➪Workbook pages 58/59

Vocabulary and speaking

1   Label the vertical axis, the lateral axis and the longitudinal axis on the diagram.

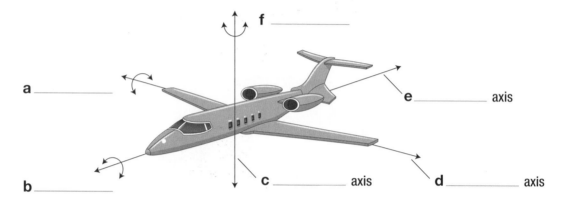

f _____

a _____

e _____ axis

c _____ axis

d _____ axis

b _____

2   Read these definitions of aeroplane movements and finish labelling the diagram.

pitch: changes in pitch raise and lower the nose of the plane

roll: rotation along the longitudinal axis of the fuselage

yaw: horizontal rotation around the vertical axis of the plane

Listening and writing

Martin is 17 and has just had his first flying lesson, which was a surprise birthday present from his father. You are going to hear an instructor describing how the aircraft controls work.

1   🎧 Listen and write a list of the controls that are mentioned.
    Check you understand what they are.

2   Look at Skills Box 1. With a partner, try to complete the notes in the table on the next page.

    🎧 Then listen again to check your answers.

> ### Skills Box 1
> **Form-filling**
>
> In aviation engineering, there are a lot of forms which must be completed in note form. Remember that:
>
> - you should not use full sentences; *a(n)*, *I* and *the* can be left out;
> - you should use capital letters for clarity.

	where on aircraft?	determine(s) which movement?	controlled how by pilot?
elevators			*joystick pulled/ pushed*
ailerons		*rolling movement of plane (banking)*	
rudder			

Speaking

1   **Look at Skills Box 2. Practise saying the following phrases, which contain consonant clusters.**

against the law

the instructor was there

flight control

called the joystick

in the longitudinal axis

to increase or decrease your flying height

The flaps are near the fuselage

I want the plane to roll or bank

not exactly

pre-flight checks

> **Skills Box 2**
>
> **Consonant clusters**
>
> English words can have two or three consonant sounds at the beginning, and even four at the end.
>
> For example:
> *price, glass, spring, split*
> *next, arranged, months*
>
> These consonant clusters are important for understanding and speaking. They must be pronounced with no extra vowel sounds between them.

2   **Find the phrases in the tapescript and practise the whole sentences. Don't forget to use the right sentence stress.**

Vocabulary and speaking

1   **In the tapescript, underline five or six key phrases for explaining how the control surfaces work. Then list them on a separate piece of paper.**

2   **Practise two conversations. You must use all the phrases on your list.**

   a Between Martin's father and friend: Martin's father is telling the friend about Martin's first lesson and the things that he learned.

   b Between the instructor, Dennis Saunders, and a colleague who asks him how the first lesson went and what Martin learned.

⇨**Workbook pages 60/61**

Lesson 9
From pilot to control surface

In mechanical aircraft control systems, there are three main types of linkage between the pilot's controls and the moveable control surfaces.

Vocabulary

1 **Match the words to their meaning.**

a	tube (*n*)		**i**	hard to move; the opposite of *slack*
b	rod (*n*)		**ii**	a system of cables and wheels
c	cable (*n*)		**iii**	difficult or impossible for a person to get into
d	hollow (*adj*)		**iv**	a rope made from wires twisted together
e	rigid (*adj*)		**v**	with nothing inside; the opposite of *solid*
f	tension (*n*)		**vi**	does not bend
g	play (*n*)		**vii**	a cylinder or pipe
h	pulley (*n*)		**viii**	small amount of movement in a connection
i	inaccessible (*adj*)		**ix**	pulling force
j	stiff (*adj*)		**x**	a metal (or wooden) bar

2 **Make sure that your pronunciation of the words is correct, including word stress.**

3 **Which of the words are shown in the diagrams?**

a   b   c   d

Reading and speaking

1 **Work in groups of three: A, B and C.**

Student A: Read Text A (Torque tubes).
Student B: Read Text B (Push-pull rods).
Student C: Read Text C (Cables).

Make notes about your text using these headings.

a	material	**d**	how it works
b	shape	**e**	possible problems
c	what it operates		

Text A

Torque tubes

These hollow aluminium alloy tubes are used to transmit a turning force or torque from the cockpit controls to control surfaces. A hollow tube is, of course, much lighter than a
5   solid rod, although under excessive loads the tube could risk twisting. Torque tubes are often used to move wing flaps. The pilot uses the control column in the cockpit to turn the tubes in the opposite direction.

Text B

Push-pull rods

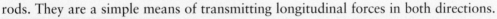

Elevators, some ailerons and helicopter rotor controls are operated by these rigid rods. They are a simple means of transmitting longitudinal forces in both directions.
5   As long as they are correctly adjusted, there should be a minimum of play in the connections at each end. They are also made from hollow aluminium tubes, to reduce weight. However, this means that, when they are used to push, there is a risk that they may bend or buckle if the load is excessive.

Text C

Cables

Because of their strength, flexibility and versatility, steel cables are the most commonly used linkages in primary control systems. By being run over
5   pulley wheels, they can change direction and be installed in small inaccessible areas of the aircraft. They are used for other parts, such as engine and emergency landing gear controls as well as the primary control system. In order to function
10 properly, the cables must be adjusted to the correct tension. Too much tension results in stiff controls and wear or damage to the pulleys; if there is insufficient tension, the controls are slack and there may be a risk that the cable will come out of a
15 pulley groove. At the ends of the cables there are special terminals which must be stronger than the cable itself. The cable should break before the terminal does, since a cable is much easier and cheaper to replace than a terminal.

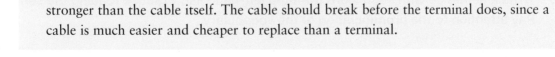

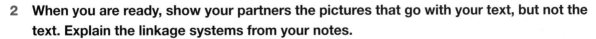

2   **When you are ready, show your partners the pictures that go with your text, but not the text. Explain the linkage systems from your notes.**

➡**Workbook pages 61/62**

Manoeuvres: Review

Vocabulary

1 Look at the Skills Box. How are these words pronounced?

a Match each word to a rhyming word.

pitch	tank
loop	head
roll	learn
turn	five
climb	make
dive	weight
bank	hole
break	which
straight	group
spread	time

Skills Box

Long vowels

English has long and short vowel sounds. The difference between them is important for meaning and understanding.

For example:

short vowels	long vowels
put	move
hit	heat

You should practise saying long vowels, e.g., /u:/ and /i:/, and vowels that consist of two sounds, e.g. /ei/ and /ai/.

b Write one more word which has the same vowel sound as each of the words on the left.

c Which vowels are short? Which are long? Which consist of two sounds?

2 Check that you remember the meaning of all the words.

3 Write a sentence with each word from the left-hand list. Practise reading the sentence. Pay attention to the pronunciation of the vocabulary.

Speaking

Look at these sketches of some manoeuvres that the team in the photo above perform. With a partner, discuss what the aircraft do to make each pattern. You will need the words you practised in Exercise 3 above.

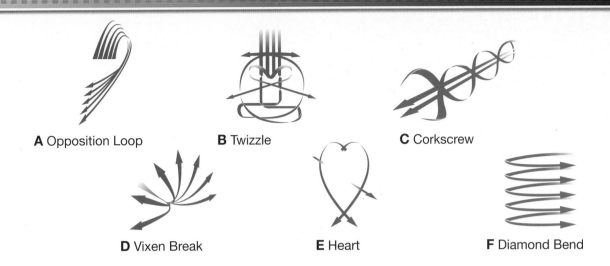

A Opposition Loop   **B** Twizzle   **C** Corkscrew

D Vixen Break   **E** Heart   **F** Diamond Bend

Listening

1 🎧 **Listen to part of a radio commentary about the British Royal Air Force Red Arrows display team. In which order are the manoeuvres A–F mentioned?**

2 🎧 **Listen again. Choose words and phrases from the lists that are used to describe each manoeuvre.**

    a   horizontally, climb, turn together, bank

    b   climb, cross, roll, bank, dive towards each other

    c   roll in long horizontal loops, up, straight and flat, up and round, inside, upwards

    d   climb, turn, loop, dive down vertically, spin

    e   horizontally, close together, break apart, spread out

    f   turn together, bank, accelerating downwards, spread apart

Writing and speaking

Work with a partner.

1   **Go back to illustrations A–F. Describe in your own words how each manoeuvre is performed, using the expressions from Listening Exercise 2.**

2   **Write notes to describe, moment-by-moment, the pilot's actions in the cockpit and the aircraft's movements for a single manoeuvre.**
    For example:

    Pilot presses r/h rudder pedal + moves stick to right. Aircraft banks ...

3   **Using the language from the lesson, discuss other manoeuvres which a group of aircraft could perform. Produce sketches/diagrams to show them.**

4   **Explain to other colleagues the manoeuvres you have discussed, using your drawings.**

    ⇨ **Workbook pages 63–66**

Speaking and vocabulary

How much do you know about car engines?

1   With a partner, discuss the difference between an *assembly*, a *sub-assembly* and a *component*. Complete the examples with the words in the box.

> spring   carburettor   internal combustion engine

assembly, e.g. _____   sub-assembly, e.g. _____   component, e.g. _____

2   Look at this exploded cutaway of an old-fashioned engine. Label as many of the numbered parts as you can.

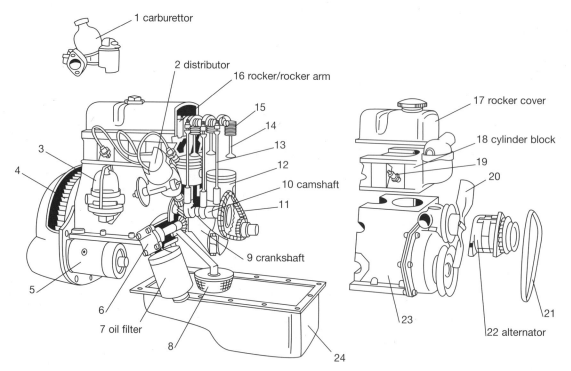

1 carburettor
2 distributor
16 rocker/rocker arm
15
14
13
12
10 camshaft
11
17 rocker cover
18 cylinder block
19
20
3
4
5
6
7 oil filter
8
9 crankshaft
23
24
22 alternator
21

3   Look at the following list of parts. How many of them can you identify immediately in the diagram? Tick (✓) the ones you know.

____ timing chain/belt   ____ spark plug   ____ oil pump   ____ flywheel
____ starter motor   ____ fan   ____ oil strainer   ____ fan belt
____ fuel pump   ____ pushrod   ____ valve   ____ engine block
____ piston   ____ sump   ____ valve springs

4   Check your answers with another partner. Use a technical dictionary if necessary to match the remaining names and numbers.

Reading

1   **Read the descriptions of parts 3, 9, 16, 19, 22 and 24 below. Find them again on the diagrams.**

2   **Work with a partner.**

   a   Complete the parts of the table which are easy for you.

   b   Use the descriptions which you have completed to help you study the diagram and complete the rest.

Part number	Description	Name
	Small chamber which mixes fuel and air to be burnt in the cylinders	
	Flexible chain or belt which connects the crankshaft and camshaft	
	Motor which turns the flywheel to start the engine	
3	Pump which supplies fuel to the carburettor	fuel pump
	Cylindrical component which moves up and down inside the cylinder	
9	The main shaft which converts linear to rotary motion	crankshaft
	Thin rod which pushes a rocker arm up	
	Replaceable filter which removes dirt and particles from the lubricating oil	
	Metal cover which protects the valves and rocker arms	
19	Small plug screwed into the cylinder which ignites the fuel mixture	spark plug
22	Belt-driven AC generator powered by the engine	alternator
24	Large container which stores the lubricating oil at the bottom of the engine	sump
	Small shaft with cams that drive the pushrods up	
	Largest part of the engine, which houses the cylinders, pistons and crankshaft	
	Small pump which circulates lubricating oil through the engine system	
16	Small arm which rocks and moves the valves up and down	rocker/rocker arm
	Strainer inside the sump	
	Small component which allows fuel mixture to enter, and hot gas to leave, the cylinder	
	Switch which sends electricity to the spark plugs in turn	
	Small springs attached to the rocker arms	
	Rotating blades which help to cool the engine	
	Top section of the engine which houses the top of the cylinders	
	Heavy wheel which maintains the momentum of the crankshaft	
	Belt which drives the alternator and fan from the crankshaft	

➡ **Workbook pages 67–69**

Lesson 2
Aero engines

Vocabulary

Here are some factors which affect the design of an engine.

> size   simplicity   power   price   reliability   weight   beauty

Work with a partner. Write down an adjective that could be used when talking about each factor and then write its opposite, e.g.:

simplicity: simple – complex

Speaking

Which of the factors above do you feel are most important in an aero engine? Compare your ideas in groups.

Listening

You are going to listen to the following conversation between Roger, a new member of the aeroplane club, asking Tom, who has been a member for a long time, his opinion about different engine types.

Listen to the first part of the conversation.

a In what order does Tom mention the factors in the Vocabulary section above? (He mentions only five of them.)

b Listen again. What does Tom say about each factor? Look at the table below and mark the statement which is not true, according to what Tom says.

size	i	____ Smaller engines have more drag.
	ii	____ A streamlined engine is better than one which is not streamlined.
weight	i	____ Less power is needed if the engine is light.
	ii	____ In the old days, engines looked better.
	iii	____ New materials make modern engines lighter.
simplicity	i	____ Simple engines can actually be difficult to build.
	ii	____ Spare parts are usually more expensive for simple engines.
		____ Some spare parts have to come from other countries.
reliability	i	____ If there's a problem, a pilot cannot stop flying.
	ii	____ Aircraft engines run at 65–75% power.
		____ A car uses 20% of the power of an aircraft.
power	i	____ Power solves the problem of air resistance.
	ii	____ Extra power can cause dangerous situations.

Listening and speaking

🎧 **You are going to listen to the second part of the conversation.**

a Look at the five engines below. Tick the three engines that Tom has experience on.

b 🎧 Listen again. Write the number of the engine(s) next to each of these advantages and disadvantages.

advantages	disadvantages
It's streamlined.	It needs a lot of spare parts.
It's powerful.	It's expensive.
It's simple.	It isn't very powerful.
It runs smoothly.	It's complicated.

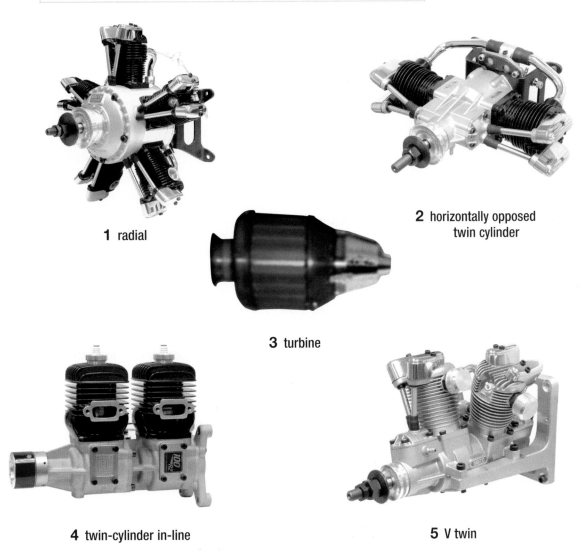

1 radial

2 horizontally opposed twin cylinder

3 turbine

4 twin-cylinder in-line

5 V twin

➡ **Workbook pages 69/70**

Lesson 3
Momentum

gyroscope

skater

top

Reading and speaking

You are going to read a text connected with these three pictures.

1 **With a partner, discuss what these three things have in common.**

2 **Look at the Skills Box.**

3 **Look at the topic sentences *a–d*. What information do you expect to find in each paragraph? With a partner, think of as many possibilities as you can.**

   a This fact was used to provide the answer to an old and difficult problem.

   b The answer is momentum.

   c Think of the skater in mathematical terms.

   d Have you ever watched an ice-skater on TV?

> ### Skills Box
> #### Topic sentences
> The paragraphs of a text often begin with the main idea of the paragraph. After it come explanations and examples.
>
> So focusing on the topic sentences allows you to read a text very quickly to get the main ideas, and also to understand better the explanations and examples given.

4 **Look quickly at the text below and match one of the topic sentences to each paragraph.**

5 **Read the text more carefully. Were your predictions in Exercise 3 correct?**

How a toy helped solve an aviation teaser

_____ If you have, you will know that one of the strangest things to see is the way that the skater can change his speed of rotation just by moving his arms or one leg. He doesn't use brakes to slow
5 down or extra power to speed up. How is this possible?

_____ The key fact to remember is the rule that momentum, which is mass x velocity (distance ÷ time in a specific direction), stays the same – you can't change it without using some kind of external force. This became important for aircraft engine design: what is true of momentum in a straight line is also true of a rotary or angular situation – momentum is stable.

_____ (See Fig. 1). His hands and leg (M) are rotating at a velocity, V – they travel a specific distance around the circumference of a circle in a specific time. They have a certain momentum. If he pulls his arms in towards his body, he decreases the radius (r) of this circle, and so the distance travelled by his hands is smaller. But since the overall momentum must remain the same, V must become higher, i.e., he spins faster. The unchanging momentum makes this spinning state a very stable one – the skater and the top won't fall over and the gyroscope will stay where you put it so long as the rotation continues.

From the very beginning of powered flight, one major difficulty was identified: vibration. The earliest engines were in-line piston engines, similar to those used in cars, which produce a lot of linear vibration (up and down, perpendicular to the axis). This put a strain on the airframe and could only be smoothed out by using a heavy flywheel to provide angular momentum. The problem was overcome by building a radial engine – with the combustion cylinders arranged in a circle, the vibration was evenly distributed and the engine more stable, like the ice-skater. The engine ran much more smoothly and put much less stress on the airframe.

Figure 1

angular momentum
$= m \times v \times r$

Speaking and writing

1   **Put the following words into the correct column depending on the pronunciation of the underlined sound. Then practise saying them.**

momentum   velocity   turn   spin   stable   radial   mass   radius
problem   vibration   circle   rotation   skater   top   distance   stress

end	stop	back	big	make	learn

2   **Look at page 247 and complete the skeleton notes.**

3   **Ask and answer questions about momentum.**

➪ Workbook pages 71/72

Why don't we turn the whole engine?

Speaking and vocabulary

1   Work with a partner. Discuss what is shown in this series of four pictures.

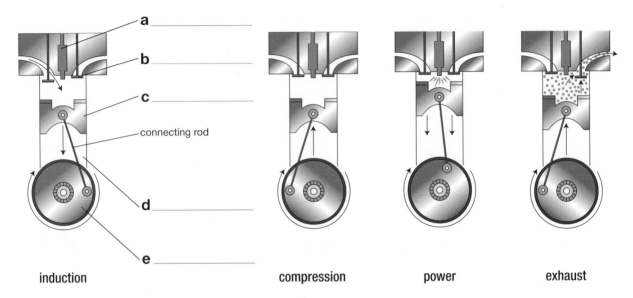

induction   compression   power   exhaust

2   With your partner, mark the words you already know in the two lists below.

the exhaust valve	opens
the ignited gas	expands
the fuel/air mixture	is sucked in
the inlet valve	closes
a vacuum	is compressed
the power stroke	is created
	begins
	is admitted

3   Match the phrases on the left with the verbs on the right.

4   Complete the labels on the first picture with the words below.

    flywheel   valve   sparkplug   cylinder   piston

Reading and writing

1 **Number the following stages of the engine cycle in order, according to the pictures.**

a ☐ The power stroke now begins – the piston is forced downwards by the pressure of the expanding gases; the exhaust valve opens.

b ☐ As the piston moves to the bottom of the cylinder, vacuum is lost and the inlet valve closes.

c ☐ The ignition spark happens before the cylinder reaches the top-centre position.

d ☐ The fuel/air mixture is sucked in by the vacuum as the piston moves downwards to the centre.

e ☐ The piston then moves upwards, compressing the fuel/air mixture.

2 **Work with a new partner. You are each going to read one text and then share the information you find.**

Student A: Read Text A on page 248 and make notes on the advantages of the rotary engine.

Student B: Read Text B on page 249 and make notes on the disadvantages of the rotary engine.

3 **Read your notes to yourself and then re-read your text (do not try to memorise it).**

4 **Look at the Skills Box. Student A should tell Student B about his text, using his notes to help. Student B should make his own notes on the talk. Then it is Student B's turn to talk.**

5 **Look at the text your partner was telling you about and compare your notes with it. Is the information the same?**

> **Skills Box**
>
> **Note-taking**
>
> When taking notes, you do not need to worry about the exact words the speaker used to give the information. What is important is that you note what the information was.

Rotary engine

Sopwith Camel (c. 1918)

➡ Workbook pages 72–74

Lesson 5
Power in numbers

Speaking and vocabulary

Look back at the diagram of an engine in Lesson 1 of this unit. How many of the parts of the engine can you see in the picture below?

Reading

Skills Box

Reading data

When you read tables containing a lot of data, you need to do two things:

- Read quickly to find a number or piece of information.
- Read more slowly to understand how the machine works.

You do not need to understand every word for these two activities.

1 Look quickly at the text below. Is it:

   a a list of problems with an engine?

   b the specification of an engine?

   c a breakdown of the design of an aircraft?

type	Merlin MK24 60° V12, pressure liquid cooled, 4-stroke, 48 valves, supercharged, geared drive
construction	Two-piece aluminium alloy crankcase, two-piece aluminium alloy cylinder blocks with steel liners, two inlet valves and two sodium-cooled exhaust valves per cylinder, operated by single overhead camshaft per bank Forged steel crankshaft with marine type fork and blade connecting rods, seven main bearings, forged aluminium alloy pistons Spur reduction gear 0.42:1
supercharger	10.2" dia. gear driven two-speed, single-stage Ratios 8.15:1 and 9.49:1

fuel system	SU AVT40 2 barrel updraught carburettor with automatic mixture and boost control. 1 Rolls-Royce-SU SUX–601 variable-stroke fuel-injection pump, discharging into eye of the supercharger Automatic fuel/air ratio control
ignition	Twin magneto: BTH C6SE12 or Rotax NSE12. 14 mm x 14 mm spark plugs with screened cabling
lubrication	Dry sump, 60 psi main pressure feed
starter	BTH CA4750, and provision for hand turning on wheelcase
bore	5.4", 137 mm
stroke	6.0", 152 mm
displacement	1,649 cu in, 27 litres
compression ratio	6:1
width	29.8", 757 mm
height	43", 1,092 mm
length	71", 1,803 mm
weight	1,540 lb, 698 kg
specific output	1 hp/cu in., 60.7 hp/litre
max piston speed	3,000 ft/min, 15.2 m/sec
power ratings	@ take-off : 1,610 hp/3,000 rpm/18 lb boost low gear: 1,635 hp/3,000 rpm/2,250 ft high gear: 1,510 hp/3,000 rpm/9,250 ft cruise: 1,015 hp/2,650 rpm/15,500 ft
fuel	100/130 grade aviation fuel max consumption @ take-off (Merlin 24) 150 gallons p/hour (3 gallons per minute), reduced to approx half this when cruising

2 **Find the following in the text as quickly as possible.**

weight in kg   dimensions of the spark plugs   material for the crankcase
fuel type   number of valves

3 **Working quickly, find and circle these numbers. What do they refer to?**

15.2 m/sec   0.42:1   SU SUX–601   BTH CA4750   757 mm

4 **Look at the Skills Box. Now take your time to answer these questions.**

a What are the cylinder liners made of?

b How long is the engine?

c How much fuel does the engine use when the aircraft is cruising?

d What is the diameter of the supercharger?

e Does the pilot control the fuel/air mixture?

➡ Workbook pages 74/75

Rebuilding an engine

Speaking

Look at the photo below. Don't read the text, but look at the headline above the photo.

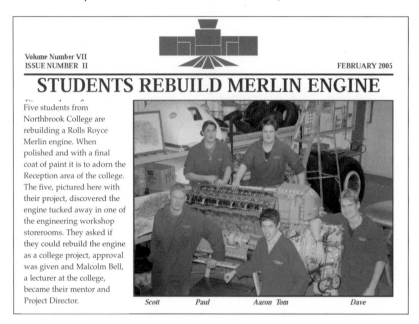

Volume Number VII
ISSUE NUMBER II

FEBRUARY 2005

STUDENTS REBUILD MERLIN ENGINE

Five students from Northbrook College are rebuilding a Rolls Royce Merlin engine. When polished and with a final coat of paint it is to adorn the Reception area of the college. The five, pictured here with their project, discovered the engine tucked away in one of the engineering workshop storerooms. They asked if they could rebuild the engine as a college project, approval was given and Malcolm Bell, a lecturer at the college, became their mentor and Project Director.

Scott   Paul   Aaron Tom   Dave

1   **With a partner, discuss these questions.**

   a What kind of place is it?

   b Who are the five people?

   c What have they done?

2   **Read the text next to the picture to check your ideas.**

Reading and writing

1   **Look at the Skills Box. Match each sentence *a–e* with one of the student's check sheet notes opposite.**

   a ☐ Scott stripped down the carburettor and cleaned it.

   b ☐ We painted the mounting display frame.

   c ☐ The cylinder heads were sandblasted.

   d ☐ The engine was moved from the storeroom to the workshop.

   e ☐ Both cylinder heads were put back on the engine.

Skills Box

Writing notes for a check sheet

Leave out *a*/*the* and the verb *be* in past passives in the same way as you do in the present passive, e.g.:
The pistons were removed.
– PISTONS REMOVED

A new chain was fitted.
– NEW CHAIN FITTED

Check sheet notes

i Engine moved from storeroom to workshop ☐

ii Supercharger removed and stripped down ☐

iii Carburettor taken off supercharger ☐

iv Carburettor stripped down and cleaned ☐

v Carburettor reattached to supercharger ☐

vi Cylinder heads removed ☐

vii Pistons removed and cleaned ☐

viii Cylinder heads sandblasted ☐

ix Pistons and cylinders refitted ☐

x Cylinder heads reattached ☐

xi Supercharger reassembled ☐

xii Supercharger remounted onto engine ☐

xiii Mounting display frame painted ☐

xiv Engine transferred to display position ☐

2 **In your notebook, write full sentences for the rest of the check sheet notes.**

Listening

Now listen to someone who rebuilt a working Merlin engine.

1 🎧 **Listen once for this general information.**

a Where did the engine come from?

b Did he have to do a lot or only a little work on it?

c Is it working OK now?

2 🎧 **Listen again. Tick the six problems on the left which you hear.**

	Problems	Solutions
a	☐ water leaking	carb. stripped, inspected, reassembled
b	☐ oil leaking	valve dismantled and cleaned
c	☐ spark plugs dirty	spares bought and fitted
d	☐ carburettor flooding	brass tubes made, old ones replaced
e	☐ bolts loose	plugs cleaned and refitted
f	☐ pistons worn	new cylinder head gasket fitted
g	☐ valve/valve springs damaged	front blocks machined to fit
h	☐ exhaust valve blocked	engine stripped down completely
i	☐ cylinder head leaking again	all fastenings checked
j	☐ cylinder blocks out of alignment	valve and all springs replaced

3 🎧 **Listen once more. Match a solution on the right to each problem.**

➯**Workbook pages 75/76**

Lesson 7
A new idea

Vocabulary and speaking

1 Complete the words for parts of an engine by filling in the missing letters.

   a p__p_ll_r   **c** exh__st   **e** sh__t   **g** t_r__ne   **i** in__ke
   b g__rb_x   **d** c__b_s_i_n   **f** ch_m__r   **h** c__pr_ss_r

2 This is a cutaway diagram of a gas turbine engine. Work with a partner to put the words from Exercise 1 in the right places.

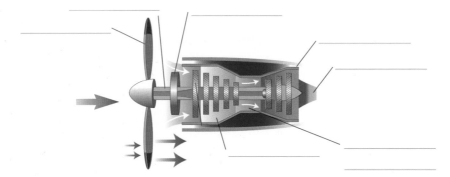

3 Look at the path of the arrows through the engine. Discuss with your partner what it shows. Use these expressions.

air goes into

the compressor acts on the air

the air is mixed with fuel

the hot gases drive the turbine

push the turbine blades

the propeller uses the power

the exhaust gases go out

Reading

1 Read the text and see if your ideas in Exercise 3 above were right.

> The turboprop engine
>
> The next stage in the design of aircraft engines was the development of the turboprop engine. This engine is a type of gas turbine which has a propeller very similar to the ones that are used by piston engines, but which is driven by the combustion of gas in
> 5 a single combustion chamber instead of several cylinders. Turboprop engines are usually fitted to small or medium-sized aircraft where speed is not the primary requirement.

The main components of a turboprop engine are the intake, the compressor, the combustion chamber and the turbine. Air is drawn into the intake and compressed by the compressor. Fuel is then added to the compressed air in the combustion chamber and is ignited by a spark. The hot combustion gases expand through the turbine, to provide power to the turbine by exerting pressure on the blades, causing the central shaft to rotate. Some of this rotary power drives the compressor, and the propeller is driven by the remaining power via a reduction gearbox. In some turboprop designs, the exhaust gases are expelled directly from the rear and can provide additional thrust.

2 Underline an expression in the text which has a similar meaning to each of the expressions in Exercise 3.

Language

1 Complete each of the following phrases with a preposition.

a draw something _____

b add something _____

c provide power _____ something

d exert pressure _____ something

e expel something _____ somewhere

> ### Language Box
> *by*
>
> In passive sentences, the person or thing which does the action is sometimes mentioned. If so, the preposition *by* is used.
>
> For example:
> *the propeller is driven <u>by the remaining power</u>*
>
> *similar to the ones that are used <u>by piston engines</u>*

2 Look at the Language Box. How many examples of *by* can you find in the text? Are they all passive sentences?

Speaking

1 Look at the diagram below of the Pratt & Whitney PT6 engine. Look carefully at the labels: *propeller shaft, reduction gearbox,* etc. Discuss the following questions with a partner.

a Where is the air intake?

b What is the path of the air through the engine?

c Are there any other differences between this engine and the description in the text?

2 Work with your partner to test him on his knowledge of this engine. Ask and answer questions like:

What does the ... do?

What happens after ... ?

Where is the ... ?

⇨ Workbook pages 76/77

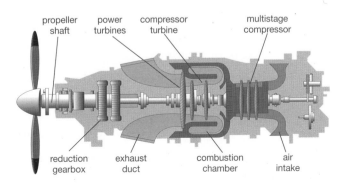

Turbo stats

Speaking and reading

1 Talk to a partner and decide whether these things might be true or false, and why.

   a When a propeller rotates faster, the plane always goes faster.

   b When you travel higher, the air is thinner.

   c Most modern aircraft do not have propellers.

2 Read the text to check if you were right.

Jet engines

Although the turboprop engine is still widely used, there are certain limits that it
cannot exceed. Propellers become less and less efficient as they rotate faster, until the
point where little or no extra aircraft speed can be achieved. They are also less
5 efficient as the altitude increases, because the air becomes thinner. The solution to this
problem is the jet engine. In its basic form it is known as a turbojet, and it was this
type of engine which was used in the first jet fighters and airliners.

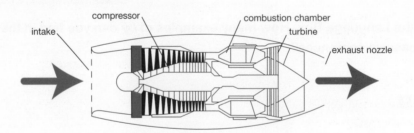

Cross-section of a turbofan engine

However, most modern aircraft turbine engines are 'turbofan' engines. Like the
turboprop engine, a turbofan engine uses a turbojet as a central core; the difference is
10 that it has a large fan mounted in front of the compressor section – the fan is clearly
visible in the front of a commercial airline jet engine. The fan and the compressors are
driven by a turbine in the same way, but here the fan acts as the propeller: it pushes
air to create thrust. It differs from a conventional propeller in that it has a lot of wide
blades spaced closely together and is surrounded by a tight cowling. The turbofan
15 looks similar to a turbojet and is more or less the same size, but the addition of the
fan at the front makes the engine quieter and more efficient than the basic turbojet.

3 Label the fan on the diagram of the above.

Listening and speaking

1 **Say these numbers with a partner. Make sure your pronunciation, including stress, is correct.**

4,500	1¾
11,000,000	167
40	3.142
14	490,000
115.8	9⁄16

> ## Skills Box
> ### Numbers
> You know the numbers very well. However, it is extremely important in engineering that you hear and say them correctly, so pronunciation practice is needed.
>
> For example:
> 2½, 1⅛, ¼
> a million
> one thousand two hundred

2 **You are going to listen to a power plant engineer describing one second in the life of a Boeing 747 engine. Before you listen, look at the numbers in the left-hand column of the table below, and at the Skills Box. Practise saying the numbers with a partner. Don't worry about the right-hand column yet.**

3 **Circle the number on the left which you hear in each case. The first one is already done.**

4 **Listen again. Match the phrases on the right with the numbers on the left.**

5 **Work with a partner to see how much you can remember. Ask and answer questions about the information in the table, e.g.:**

How much air is drawn in per second?

3,500 ft (35,000 ft)	cruising altitude of Boeing 747-400
12,000 lb / 1,200 lb	air drawn into engine
700 lb/sec / 17 lb/sec	air bypassing engine core
8% / 80%	pressure of compressed air
150 psi / 115 psi	thrust from each engine when cruising
85°F / 850°F	fuel injected into compressed air
1¾ lb/sec / 1,340 lb/sec	exit velocity of exhaust gas
2,000°F / 22,000°F	rate of rotation of compressor
3,310 rpm / 3,300 rpm	rate of rotation of fan
9,500 rpm / 950 rpm	temperature of exhaust gas
1,400 ft/sec / 14,000 ft/sec	temperature of combustion
10,000°F / 1,000°F	temperature of compressed air

➡ **Workbook pages 77–79**

Lesson 9
Fuel

The easy way and the hard way to refuel

Vocabulary and speaking

Work with a partner and match the properties of fuel on the left with the descriptions on the right.

a	flash point	i	How easily an electric current flows through the fuel.
b	freezing point	ii	The temperature at which the fuel ceases to be a liquid and starts to become a solid.
c	sulphur content	iii	How easily a liquid changes to a vapour and combusts.
d	lubricity	iv	The amount of heat produced when a given quantity of fuel is burned in the engine.
e	volatility	v	The ability to make moving parts in the fuel system, such as pumps and valves, operate smoothly.
f	electrical conductivity	vi	Damage done by the fuel to metal components by chemical reaction.
g	net heat of combustion	vii	The percentage of sulphur contained in a fuel.
h	corrosiveness	viii	The temperature at which a fuel will give off a vapour that will burn instantly when ignited.

Reading

Read the text below about the properties of aviation fuels. Check your answers.

In the early days of the jet engine, many supporters of the new technology believed that almost anything, including peanut butter and olive oil, could be used as a fuel, because jet engines were more tolerant of poor-quality fuel than piston engines. However, although it is true that jet engines are much more tolerant than gasoline and
5 diesel engines, aircraft and engine fuel systems are sensitive to the chemical and

physical properties of fuel. Petrol is light, but too volatile. Kerosene is a heavier and less volatile fuel than petrol, but lighter and more volatile than diesel fuel. That is why, for the last 50 years, kerosene has been the fuel that powers the world's airlines and military fleets.

10 The properties of a jet fuel are considered carefully when the specification is being written. Some of the most important are: the flash point of the fuel – how hot it needs to be before it burns; its freezing point (temperatures are very low at high altitudes); the sulphur content, as well as the percentage of other chemicals in the fuel; lubricity, since the fuel helps to lubricate the fuel system itself; conductivity, i.e., how well the

15 fuel conducts electrical current; the net heat of combustion, i.e., the heat gained per unit of fuel burned, and corrosiveness – it is important that some fuels will cause chemical reactions which are damaging to the fuel system.

As engines have developed since the 1940s, the standard specifications for different grades of jet fuel have developed as well, so that safety, efficiency and economy are

20 maximised.

Speaking

Discuss the following questions with a partner.

   a Why is kerosene a suitable fuel for the jet engine?

   b What combinations of properties might make a fuel very safe/unsafe, and very economical/uneconomical?

Language

1 Look at the Language Box. Expand the questions using the patterns below.

over the last … years

since … date

for … years

   a Why / fuel specifications / develop / the 1940s?

   b How / fuel / become safer / 50 years?

   c What materials / invented / century?

   d How / aircraft shapes / change / 1950s?

   e five decades / what major changes (modifications) / made / to engines?

> ## Language Box
>
> **Changes over time**
>
> When you look back over a period of time, you use *have* + the past participle.
>
> For example:
> *Since the 1940s, specifications **have developed**.*
>
> *Which fuel **has powered** the world's airlines for the past 50 years?*
>
> This verb structure is often called the **present perfect tense.**

2 Discuss your answers to the questions with a partner or in small groups.

⇨ **Workbook pages 79–81**

Lesson 10
Moving the fuel around

Speaking and listening

You are going to listen to an instructor talking about fuel.

1 **Before you listen, discuss with a partner what problems there might be with fuel when a modern aircraft is in flight. Which of these do you think are a problem and why?**

  a Fuel is very heavy.

  b It might get too hot and start a fire.

  c The fuel tanks might leak.

  d The fuel might freeze.

  e It might move and change the balance of the aircraft.

  f It smells bad.

  g The pilot might not know how much he has in the tanks.

2 🎧 **Listen to the first part of the talk and see if your ideas are mentioned.**

3 **Look closely at the five pumps shown from a small aircraft. With a partner, decide which pump:**

  a is connected to a battery?

  b is just a tube?

  c has a filter before the fuel outlet?

  d has a valve to stop fuel coming back?

  e has a rotating part to push fuel through?

4 🎧 **Listen to the rest of the instructor's talk. As you listen, match the drawings with the correct titles below them, *i–v*.**

a ☐

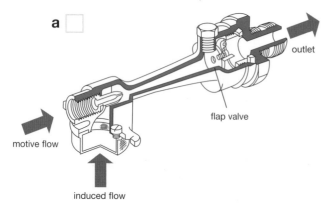

motive flow · outlet · flap valve · induced flow

b ☐

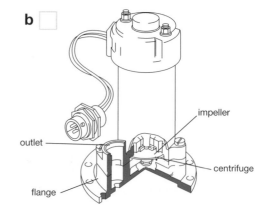

impeller · outlet · centrifuge · flange

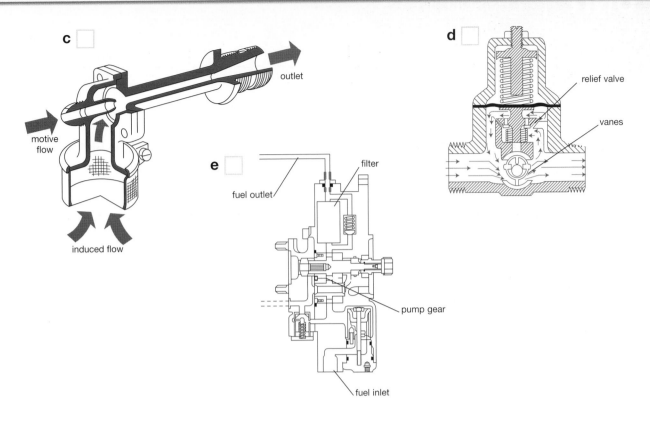

c ☐

motive flow

outlet

induced flow

e ☐

fuel outlet

filter

pump gear

fuel inlet

d ☐

relief valve

vanes

i main (rotating vane) pump   **ii** delivery jet pump   **iii** high-pressure gear pump

iv transfer jet pump   **v** electric booster pump

5 🎧 **Listen again to the second part of the recording and complete the notes.**

pump	notes
main (rotating vane) pump	
transfer jet pump	Tube. Venturi principle – sucks fuel. Transfer fuel wing tanks to collector tanks.
high-pressure gear pump	
delivery jet pump	
electric booster pump	28 V. Runs from battery to engine. 1 in each collector tank, fixed. Sealed, in liquid.

6 **Using your notes, take turns with your partner to describe the pumps to each other.**

⇨**Workbook pages 81–84**

Innovation: Review I

A good example of the innovative use of existing technology is a British car called the BMC Mini, which was designed in the 1950s and was produced right up until 2001. A newer version based on the original version is now produced by BMW.

Speaking

Talk with a partner.

How can a car manufacturer make a new product:

    a easy/quick to build?

    b cheap for the customer to buy?

Think about materials, size, fuel consumption, new technology, complicated systems, etc.

Reading

1 **Read the design specification for the Mini below. Does it contain any of the ideas you mentioned in the Speaking section?**

 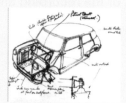 

BMC design specification, 1956

The new car must:

a be economical – petrol is very expensive mpg.

b take no more than 36 months to develop and put into production.

c be cheap to manufacture, and therefore cheap to the customer.

d seat a family of four.

e measure 3 m x 1.2 m x 1.2 m overall dimensions, of which passenger space min. 1.8 m long; remaining length for engine, gearbox and luggage.

f not use money in designing new components.

g contain a minimum of material in the structure: max. weight 650 kg.

2 Now look at the innovations which the designer, Alec Issigonis, used to solve the problems of such a tough design brief. Match each of *a–i* to one of the points in the specification on the opposite page. There may be more than one possible answer in each case.

  a He used a small engine which was already in production.

  b He did what he could to minimise the overall dimensions.

  c The engine was turned sideways.

  d The gearbox was fitted under the engine.

  e Front-wheel drive was used, so no drive shaft ran under the floor.

  f The wheels were made very small (250 mm diameter).

  g Independent rubber suspension was used to reduce structural stress.

  h The first prototype was assembled by the end of 1957; full production began in 1959.

  i The body of the car was constructed so that it was thin and light.

Vocabulary

1 These words are all verbs. Make a noun from each one.

> contain   measure   increase   design   develop   produce   connect   fit   support   assemble

2 Choose the right form, verb or noun, of the words from Exercise 1 to complete these sentences.

  a Issigonis' _____ was innovative in its use of existing technology.

  b The first Mini went into full _____ in 1959.

  c In front-wheel drive, there is no _____ between the engine and the rear wheels.

  d The wheels are _____ by independent rubber suspension.

  e The _____ of the car had to take a maximum of three years.

  f The cabin needed to _____ seating for four people.

  g Issigonis had to _____ the dimensions of the original design.

  h The passenger cabin had to _____ into 1.8 m of the overall length.

  i _____ of the prototype was complete by late 1957.

  j 3 m x 1.2 m x 1.2 m were the outside _____ .

3 Now make sentences of your own using the other forms of the words, either about the Mini or any other product.

Writing

Look back at the list *a–i* in the Reading section above. Make notes on the solutions which Issigonis brought to the difficult design brief.

⇨ Workbook pages 85/86

Reading

1 **Read the descriptions of different types of drawing. Which one is shown below?**

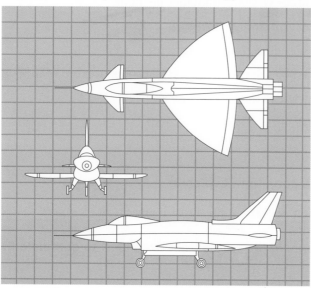

___ **Sketches**

These are rough drawings made by hand, without the use of instruments.

___ **Assembly and 'exploded' drawings**

The components are shown separated from each other to display the way they are assembled.

___ **Installation drawings**

These drawings show the location of the parts and assemblies of specific systems.

___ **Sectional drawings**

These show how a component would look if it was cut through the middle.

___ **Cutaway drawing**

A cutaway drawing shows the outside of a component with part of it cut away.

___ **Perspective or pictorial drawings**

These show the components as they look in three dimensions.

___ **Orthographic projections**

These show separate flat views of different faces of an object.

2 **Now complete each description above with one of the following sentences.**

a All the details of parts used in the installation are identified clearly. These drawings are often used by maintenance and repair technicians.

b They are often coloured and give an overall impression, to the design team or customer, of what the finished product will look like.

c They are used to quickly convey only a specific bit of information and contain a minimum amount of detail.

d These are useful for giving the overall dimensions of a product, or for a general overview of an assembly from different angles.

e A list of all the components shown is normally included, giving the reference number, part number, quantity and description of each one.

f In this way, the parts inside a larger assembly can be seen in relation to it.

g The different types of materials used in the component are shown using shading, lines and cross-hatching, and the exact point at which the component has been 'cut' is clearly indicated on the drawing.

3 **Find examples of the other types of drawing on pages 68, 82, 100, 101 and 102.**

Language: prepositions

1 **Cover the Reading section. Complete the following with on, *through*, *from*, *at*, *without* or *in*. Check your answers in the Reading section.**

   a the location of the parts _____ the completed aircraft

   b as if it was cut _____ the middle

   c the point _____ which it has been cut

   d they are separated _____ each other

   e assemblies _____ the completed aircraft

   f materials used _____ the component

   g made _____ the use of instruments

   h they can be seen _____ relation to each other

   i an overview of the assembly _____ different angles

   j it is clearly indicated _____ the drawing

   k as they look _____ 3D

2 **Find at least ten more expressions with *of* in the Reading section.**

description

view

a/the _____ of

3 **Apart from the combination in the texts, how many more can you think of for each expression, e.g.,
a list of: *parts, jobs, names, numbers, shopping.***

4 **Make sentences of your own using some of the expressions.**

➡ **Workbook pages 86/87**

Vocabulary

1 **What kind of tools could you use to:**

a grip? _____

b loosen/tighten? _____

c cut? _____

d beat? _____

e make a hole? _____

f measure? _____

g file? _____

h open? _____

i strip? _____

j turn? _____

k mark out? _____

2 **Match each screwdriver type to a picture.**

pozidrive bit   hex bit   slotted bit   Phillips bit

a _____ **b** _____ **c** _____ **d** _____

Reading

Look at the specifications for two multi-tools on the next page.

1 **Which of the attachments listed can you see in the pictures?**

2 **Read the specifications to find:**

a which is longer.

b which is lighter.

c which has a ruler.

d which has scissors.

e how many attachments they *both* have, *e.g., wire strippers.*

Leatherman	Victorinox
Length: 4 inches	Length: 91 mm
Weight: 6 ounces	Weight: 157 g
Locking pliers	Large blade
5 Wire cutters	Small blade
Hard-wire cutters	Can opener & small screwdriver (3 mm)
Serrated knife	Pressurised ballpoint pen
Wood/metal file	Stainless steel pin
Small screwdriver	Mini screwdriver 1.5 mm
10 Medium screwdriver	Bottle opener & screwdriver (6 mm) with wire stripper
Large screwdriver	Pliers
Ruler (inch/metric)	Keyring
Lanyard attachment	Tweezers
Phillips screwdriver	Toothpick
15 Hex bit driver	Scissors
Wire stripper	Reamer, punch & sewing eye
Bottle opener	Wrench with
	– 4 mm & 5 mm female hex drive
	– 4 mm posidrive 0 & 1 bits
20	– 4 mm slotted bit
	– Phillips 2 bit
	– 4 mm hex bit

Speaking

Look at the two multi-tools. Describe them.

a How is the design of each multi-tool different?

b Who might they be designed for?

c Could they be dangerous to use?

d How should you look after them?

Writing

**1 How are the attachments used? Write simple instructions.
For example:**

The wire stripper: First, decide how much of the plastic sleeve to remove. Then grip the sleeve with the wire strippers …

2 Compare and contrast the two tools. Think about the user, possible uses and positive and negative aspects. For example:

The Victorinox has more attachments, but it doesn't look as strong as the Leatherman.

⇨ Workbook pages 87/88

Tools and techniques: Review

Speaking

Look at the list of hand tools below. How could each of these cause an accident if you:

a didn't put it away properly?

b used it for the wrong job?

c didn't take care when using it?

> hammer saw shears file micrometer drill spanner

For example:

If you didn't pay attention when you were using a saw, it could cut you.

If you dropped a spanner, it could hurt someone or break something.

Listening and reading

You are going to listen to two conversations about a workshop accident.

1 🎧 **Listen to Part 1 and circle the correct answer.**

a Who are John, Frank and Mr Green?

   i friends

   ii supervisors

   iii two workers and a supervisor

b What was Frank doing?

   i drilling

   ii milling

   iii guillotining

c How did the accident happen?

   i the chuck was left in the drill

   ii the aluminium sheet fell

   iii the drill was broken

d Which part of his body did he injure?

   i his head

   ii his finger

   iii his leg

2 **Look at the form opposite.**

a What is the form?

b Look at the heading/title for each section of the form. Find the section that answers each question.

   i how the accident happened

   ii what caused it

   iii what kind of injury it was

   iv who was injured in the accident

   v who or what caused the injury

   vi which part of the body was hurt

   vii how to stop this happening again in future

   viii the accident reference number

c 🎧 Listen to Part 2 and complete the form. Listen more than once if you wish.

rm for the reporting of accidents, incidents and occupational ill health

se complete all sections and tick boxes as appropriate.

tails of injured person:

ne _____

upation _____ Age _____

le ☐ Female ☐

ployee ☐ Contractor ☐ Pupil ☐ Client/Public ☐

Date accident happened _____ Time _____
Address and tel. no. of premises where accident happened

Date notified _____ To whom _____
Names and addresses of any witnesses 1 _____
2 _____

pe of injury/ill health (give dates):

ue	/ /	Partial loss of sight	/ /
.th	/ /	Concussion	/ /
er 3-day absence	/ /	Shock	/ /
cture	/ /	Poisoning/Gassing	/ /
ocation	/ /	Internal injury	/ /
n	/ /	Hearing impairment	/ /
ld	/ /	Disease	/ /
/scratch	/ /	Irritation	/ /
cture wound	/ /	Strain/sprain	/ /
ise/swelling	/ /	Other _____	/ /
s of limb	/ /		
s of sight	/ /	First-aid treatment	/ /

Agent of injury:

None ☐
Handing/lifting ☐
Hit by moving object ☐
Hazardous substance ☐
Electricity ☐
Pressure system ☐
Machinery (powered) ☐
Heat or cold ☐
Machinery (hand-held) ☐
Animal/insect ☐
Slip, trip or fall ☐
Human ☐
Other _____ ☐

e of injury:

ad ☐   Chest ☐   Abdomen ☐   Back ☐   Internal ☐

	Left	Right		Left	Right		Left	Right
	☐	☐	Elbow	☐	☐	Finger	☐	☐
e	☐	☐	Lower arm	☐	☐	Upper leg	☐	☐
ck	☐	☐	Wrist	☐	☐	Knee	☐	☐
t	☐	☐	Hand	☐	☐	Lower leg	☐	☐
er arm	☐	☐	Ankle	☐	☐	Shoulder	☐	☐
			Toes	☐	☐	Hip	☐	☐

me and address of doctor _____

ne/Fax _____ E-mail _____

Cause of incident:

None ☐
Unsafe environment ☐
Unsafe machinery ☐
Unsafe stacking ☐
Unsafe system of work ☐
Misuse of equipment ☐
Manual handling ☐
Horseplay ☐

ief description of incident: (attach additional sheets if required)

ecommendations/observations on remedial measures to prevent recurrence: (indicate dates, etc.)

te incapacitation commenced _____ Date of return (if known) _____ Total absence _____

property/asset damage Yes ☐ No ☐   Estimated costs _____

nature of person completing form _____ Print name _____

rkplace address and occupation of person completing form _____

ereby declare that the information contained in this report is true and no material information within my knowledge in regard thereto has been withheld.

estigating officer/Supervisor signature _____ Date _____

signation _____ Tel _____

Original	To be retained locally
Duplicate	Send to Health and Safety Team

➪ **Workbook pages 89/90**

Aircraft structure: Review

Vocabulary

1 Make two lists or spidergrams under the headings 'Parts of an aircraft structure' (e.g., *wings, rudder*) and 'Flight manoeuvres' (e.g., *bank, roll*).

2 Compare your list with a partner and add to it. Discuss how each part of the aircraft structure is used in the manouevres.

Reading and writing

1 Read the text below quickly and circle all the things that are mentioned in your lists.

2 Read the text again and decide which project is shown in the picture.

> Variable geometry, in theory and in practice
>
> Being able to alter the overall shape of an aircraft's tail and the length of its wings, for example, during flight could make existing designs much more versatile and capable of complicated flight manoeuvres. DARPA, the US Defense Advanced Research Projects
> 5 Agency, has three such ideas on the drawing board.
>
> — **Project 1: Lockheed Martin is exploring an origami-like folding wing:**
> - We know that gaps in a wing joint cause drag, and that this rips a wing apart at high speeds.
> - The new wing will bend and fold without open joints.
> 10 - The plan is that 'Shape Memory Polymer' coverings will be used to seal joints: as the wing changes shape, the polymer will stretch with the movement.
> - Tiny heaters could perhaps be attached to the wing. When a shape change is complete, these would heat the SMP and so reseal the joints around the new wing shape.

15   • If a pilot could change the wing shape and size quickly enough, shortening one wing slightly and raising or lowering the other, it would be possible to turn, bank and dive much like birds do.

—   **Project 2: Raytheon is developing a telescopic wing:**

     • When this design is in operation, wing sections will move in and out of the
20   fuselage.

     • This type of shape-changing has been tried before: the US, Europe and the USSR all developed aircraft in the 1960s and 1970s that could swing their wings forward or back for take-off, diving or cruising.

     • The aim is that the wing will be able to change its surface area by more than 150%.

25   • Possible smart materials, such as SMPs, and new types of actuators, would allow extension.

—   **Project 3: NextGen Aeronautics is working on a batwing:**

     • A series of sliding skins will disconnect, extend to create new wing shapes, and then reconnect.

30   • 'Shape Memory Polymers' have existed since the 1930s. These stretch and then resolidify when heated.

     • SMPs will be used to cover joints.

     • A batwing would save weight and reduce drag, since there would be no need for ailerons, rudders or tail fins.

35   • Theoretically, if the aircraft's whole wing was turned into a reshapeable control surface, the pilot could simply twist the whole wing to climb, roll, etc.

3 Look at each bullet point in the texts and decide whether it describes:

a the possibilities of shape-changing theory (mark T).

b the new design (mark D).

c facts about existing technology (mark F).

4 Reduce the information to notes under the three headings *Facts*, *Theory*, *Design*. For example:

> Facts
> gaps ⟶ drag: v. dangerous @ high speed
> Design
> new wing ⟶ will bend and fold without open joints

Speaking

Work with a partner.
Take turns describing one of the new designs for shape-changing wings, the problems and the possible solutions. Use your notes to help you.

⇨ **Workbook pages 91/92**

Lesson 6
Stresses: Review

Vocabulary

1 **Match the two halves of the phrases.**

pull something	pulls the plane forward
a squeezing or	over another
wings are subjected	another
the force is	under excessive loads
skin panels are	resisted by structural strength
one force acts against	riveted together
one piece of material slides	to compression forces
the propeller	crushing force
parts can fail	apart

2 **Draw sketches with arrows to show the meanings of the phrases above. Compare them with your partner's. For example:**

parts can fail under
excessive loads

Reading

1 **Talk to a partner. How much do you remember about these five kinds of mechanical stress?**

> shear   torsion   bending   compression   tension

2 **Read these texts and write one of the five mechanical stresses in each space, A–E.**

A _____ stress:

- Engine power and propeller pull the aircraft forward.
- Fuselage, wings & tail section resist this force, acting against it because of airflow around them.
- Result is a stretching effect which tries to pull the airframe apart.

Solution(s): Appropriate tensile strength (measured in pounds per square inch (psi) of the airframe material & assemblies.

B _____ stress:

- Squeezing or crushing force that tries to make parts smaller.
- Aircraft wings are subjected to these stresses.

Solution(s): Anti-compression design resists the crushing force on an assembly.
Ability of material to meet this stress requirement in psi or Pascals.

C _____ stress:

- Engine power (torque) and rotating propeller try to twist the forward fuselage of an aircraft.

Solution(s): Resisted by the torsion strength of fuselage assemblies.
Aluminium able to stretch and not fail under a twisting load.
Al used extensively for the skin.

D _____ stress:

- Tends to make one piece of material slide over another.
- Skin panels on a fuselage are riveted together. This force tries to make the rivets (and bolts) fail under excessive loads.

Solution(s): Selection of rivets & fastenings with adequate resistance critical.

E _____ stress:

- Combination of compression and tension.
- The material on the inside of the bend is compressed and the outside material is stretched, e.g., g-loading the aircraft's upper-wing surfaces are subject to compression, while the lower-wing skin experiences tension loads.

Solution(s): Tensile and compression strength of materials/assemblies.

Speaking and writing

Work with a partner. Decide what kinds of forces and stresses are operating in the following day-to-day situations. How do they work? Use the vocabulary from this lesson. For example:

Squeezing water out of a cloth: Torsion is operating.
The two ends of the cloth are turned in opposite directions.

a picture hung on a nail in a wall

using a screwdriver

pulling out a nail using pliers

a line with clothes hung on it

a tree in the wind

food between your teeth

the head of a bolt when it is fitted

a bicycle chain

a washer under a bolt

the material in a curtain

a road surface

screws on a door hinge

⇨ **Workbook pages 92/93**

Controls: Review

Reading

Read the explanation of an aircraft braking system and label the diagram using the words from the box.

Brakes

In principle, the braking system of a small aeroplane is like that of a car.

a The system functions using pressurised hydraulic fluid.

b The hydraulic fluid is stored in a special container called a fluid reservoir.

c Each side of the system has a large cylinder, called the master cylinder.

d The brake pedal in the cockpit is connected to the master cylinder.

e When the pilot presses down the brake pedal, the force is transmitted to the master cylinder via a system of rods and levers.

f This operates the piston in the master cylinder so that pressurised hydraulic fluid is forced through a tube to smaller cylinders.

g The smaller cylinders are called the slave cylinders.

h The slave cylinders are fitted to the brake mechanism in the wheel assembly.

i The piston in the slave cylinder is smaller than the one in the master cylinder, so the original force on the brake pedal is multiplied.

> slave cylinder   brake pedal   lever   fluid reservoir
> master cylinder   tube   rod   brake mechanism   wheel

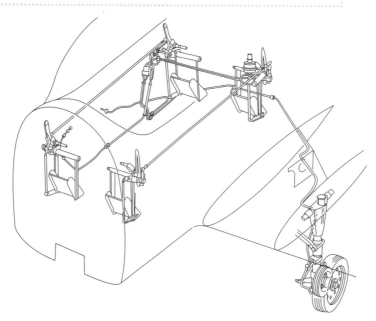

Vocabulary

1 **Look again at the text and check you understand the meanings of the verbs in Column 1 in the table below. Match them with a verb with a similar meaning in Column 2.**

1	2
operate	install
connect	work
transmit	make something work
force	send
fit	join
function	push

2 **Look at the verbs from Column 1, underlined in the sentences below. Five of them are used correctly and five of them are wrong.**

a Put a tick (✓) beside the correct sentences and a cross (✗) beside the incorrect ones.

b Rewrite the five incorrect sentences below, replacing the verbs with more suitable ones.

   i If an assembly is broken, a technician <u>operates</u> it so that it works correctly.

   ii Hydraulic systems <u>function</u> using fluid under pressure.

   iii Large assemblies are <u>transmitted</u> by air and by train to the manufacturing plant.

   iv The sump <u>connects</u> the oil filter.

   v The piston is <u>forced</u> out of the chamber by the fluid.

   vi Power from the power plant is <u>transmitted</u> to the propeller via the crankshaft.

   vii When the electrical systems have been <u>fitted</u>, they are thoroughly tested.

  viii Before take-off, the aircraft's fuel tanks are <u>fitted</u> with kerosene.

   ix The wings are <u>connected</u> to the spar by bolts.

   x In normal flight, the pilot <u>forces</u> the control stick from side to side and forwards and backwards.

Writing

1 **Make correct sentences using the verbs from Column 2 in Exercise 1 above:** *join, send, push, install, work.*

2 **Look again at the explanation of a braking system in the Reading section opposite. Choose another control system, *flaps* or *rudder*. Write a numbered explanation of how it operates, including the pilot's actions, linkages, hydraulics and the effect of that control surface on the aircraft's flight. Look back at Unit 4 for help if you wish.**

➡ **Workbook pages 94/95**

Speaking and vocabulary

1 With a partner, make sure that you know the pronunciation and stress of these words.

> check/control   replace/repair   damage/break   inspect/open
> leak/flood   pressure/heat   empty/dismantle   refill/refit

2 These are all words you have seen and used before. What is the difference in meaning between each pair?

Listening and writing

You are going to listen to a telephone conversation between a customer and a mechanic at a garage. The customer is enquiring about work done on the brakes of his car.

CALIPER ASSEMBLY
WHEEL BEARINGS
BRAKE PADS
WHEEL STUDS
DISC ROTOR

1 🎧 Number the problems with the brake system in the order that you hear them.

a		RIH FRONT DISC DAMAGED + OUT OF ALIGNMENT
b		BRAKE MASTER CYLINDER LEAKING
c		LEAK IN FLUID LINE LIH FRONT BRAKE

2 Write the problems in the job sheet opposite under 'fault'.

3 🎧 Listen again. What is the cause of each fault? Choose from this list and then add them to the job sheet (in capital letters).

		EXCESSIVE HEAT DUE TO OVERWORK – LIH BRAKE AT LOW PRESSURE
		DAMAGED TUBE
		BRAKE FLUID ESCAPING. AIR GETTING IN

4 🎧 Listen once more and add to the job sheet the corrective action taken, i.e., what the mechanics have done to solve the problems. Remember to use the standard note form, e.g., SPARK PLUGS CLEANED AND REFITTED.

You will need these verbs: *replace, empty, refill, machine, straighten, check*. Make sure you remember their meaning before you listen.

JOB SHEET: 20,000-MILE SERVICE

fault	cause	corrective action
1		
2		
3		

Vocabulary

🎧 **Look at this section from the telephone conversation you listened to. Choose which of the two words in each case fills the gap. Then listen and check your answers.**

Ted: There were a couple of things, actually. First of all, there was a <u>leak / flood</u> in the master cylinder; brake fluid was getting out and air was getting in.

Mr A: So did you repair it?

Ted: Oh no, now it would be expensive. No, it's cheaper to <u>replace / repair</u> them. So
5   we did that … we also found a slight leak in the fluid line to the left-hand front brake. So obviously we had to <u>empty / dismantle</u> all the fluid out of the system to replace the damaged tube.

Mr A: And you <u>refilled / refitted</u> it with new fluid.

Ted: Yes, and I'm afraid that we also found the surface of the disc on the right-front
10   wheel was <u>damaged / broken</u> and it was slightly out of alignment.

Mr A: Why was that?

Ted: Hard to tell. It's usually caused by excessive heat, you know, if the brakes are having to do too much work … When we <u>inspected / opened</u> it, it looked as if – have you been doing any hard or fast driving recently?

15 **Mr A:** Yes, we spent the week before last driving in the mountains. That's when I noticed there was vibration and the car was pulling to one side when I pressed the brake pedal. You have to do a lot of braking on those roads.

Ted: Sure. Yes, I should think that front wheel brake unit was doing most of the work because the other brake wasn't getting sufficient <u>pressure / heat</u>. Anyway,
20   it's all fine now. We machined the damaged disc surface and straightened it again. We've <u>checked / controlled</u> the other three discs and they're all OK …

⇨**Workbook pages 95/96**

Reading and speaking

1 **Discuss the following questions in small groups.**

   a What sort of problems do vehicles have with oil filters?

   b Why is it especially important that aircraft have effective oil filters?

2 **Read the texts that describe how aircraft oil filters operate. Label one text 'Normal operation' and the other 'Bypass operation'.**

3 **Look at the diagrams on page 119 and decide which text each one relates to.**

_____

Oil from the oil pump flows into the assembly under pressure through holes at one end, lifts the check valve off its seat and flows into the filter housing.

The oil then passes through the filter element into the central core and out again from
5   the centre of the element.

Any solid matter (such as dirt or tiny pieces of metal) which were in the oil stay on the outer surface of the filter element.

When there is no oil pressure, the check valve is closed by a spring to prevent oil flowing by gravity into the engine after shutdown.

10   This also makes it possible to replace the filter element without emptying the oil tank.

_____

If the oil filter becomes blocked, a bypass for the oil is provided by a spring-loaded valve piston, installed between the check valve and the inner end of the filter element.

The bypass valve is normally closed and sealed from the centre of the element.

15   The increased oil pressure caused by the blockage compresses the bypass valve spring and moves the valve piston.

This means that oil can flow directly into the centre of the filter element.

A simple metal filter with larger holes is attached to the inner end of the filter element and this filters the bypass oil.

20   The filtered oil passes through the core and flows out into the engine as usual.

This bypass layout makes it possible for lubricating oil to continue flowing to the power plant. When the plane lands, the blocked filter can then be changed.

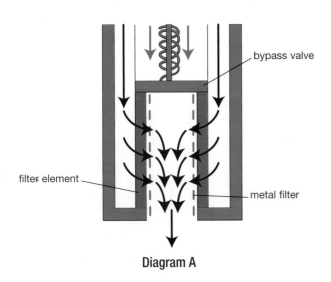

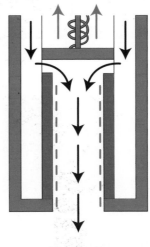

Diagram A   Diagram B

Writing

Now transfer the information in the texts/diagrams to this flow chart by putting notes in the right places. Two boxes have already been completed as examples. Notice the use of note form.

The oil enters the filter element housing

The clean oil passes into core

The clean oil then flows out of the core into the engine

~~If the filter is blocked, the oil flows into the filter through a spring-loaded bypass valve and a bypass filter~~

The oil passes through filter element and into the central core

Any solids which were in the oil stay on the outside of filter element

The oil flows into the engine in the normal way

~~The oil from the oil pump enters the filter assembly~~

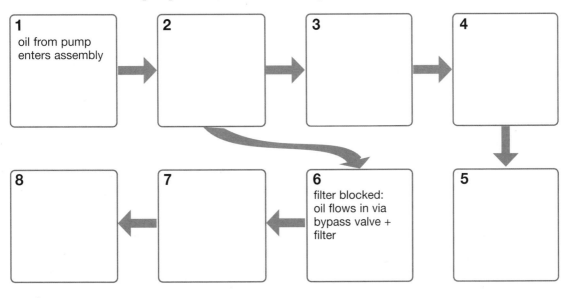

| 1 | 2 | 3 | 4 |
| oil from pump enters assembly | | | |

| 8 | 7 | 6 | 5 |
| | | filter blocked: oil flows in via bypass valve + filter | |

➡ **Workbook pages 96/97**

Engine servicing: Review

Vocabulary and speaking

Look at the box below. Discuss which of the items in the box you might have to:

a	clean	**c**	dismantle	**e**	check	**g**	strip
b	replace	**d**	inspect	**f**	fit	**h**	reassemble

a damaged selector   an engine   a reported fault   a reworked component
a new compressor   a faulty pump   a filter   a finished job
a dirty spark plug   a leaking gasket   a hydraulic tube

Reading

All aircraft engines have to be regularly serviced and overhauled. Small engines from light aircraft are often sent to specialised companies which have the experience, knowledge and equipment to do a complete overhaul quickly at a reasonable price. This information from one of these companies describes the typical overhaul procedure.

1   **Look at the five areas of work *a–e* and match a heading from this list to each one. Do not fill the gaps yet.**

New parts   Pre-assembly   Post-assembly test and final inspection
Final assembly   Electronics and computer systems   Engine stripping
Inspection and part ordering   Quality control

a _____

Engines are received, _____ down and the parts chemically cleaned prior to being _____ for condition. All components are then _____ from sub-assemblies and sent to the component overhaul section

5   for processing. All parts requiring mandatory replacement are discarded and labelled as unserviceable. (stripped, removed, inspected)

b _____

The engine is _____ to find out if it is in compliance with the manufacturer's limits, tolerances and service information requirements. All parts to be

10 _____ are ordered from authorised distributors, and a stores system ensures that the company knows exactly where each part has come from. (checked, replaced)

c _____

All new parts received are unpacked and _____ following strict quality

15 control guidelines. Where possible, these parts are disassembled for checking. Even factory-supplied cylinder kits are _____ and inspected for correct fits and tolerances prior to being _____ to each engine. (fitted, dismantled, inspected)

d _____

20 The company does not _____ engines on an assembly line. Each engine is _____ by highly experienced engine technicians and craftsmen. The engine shop foreman & the component shop foreman have many years of hands-on experience between them. (assembled, rebuild)

e _____

25 Each reassembled engine is _____ to a bench and _____ in accordance with the manufacturer's requirements. After testing, the engine is subjected to a detailed inspection to ensure that it meets the satisfaction of the senior inspector. The engine is finally immobilised, packed and _____ back to the customer. (fitted, sent, tested)

2 Now use the words in brackets after each paragraph to complete the gaps.

Speaking and grammar

1 Find and circle the following verbs in the text above: *is/are +*:

> replaced   subjected   cleaned   ordered   dismantled   sent
> fitted   stripped   removed   assembled   tested

2 Are there any other similar *is/are* + verb expressions?

3 What nouns do they refer to? For example: Text A: *are received* refers to *engines*.

4 Work with a partner. Use the verbs and vocabulary in Exercises 1 and 3 to tell each other about each area of work in the engine servicing procedure.

⇨ Workbook pages 98/99

Vocabulary and speaking

1   Look at the posters and notices around the building where you are. What are their functions?

2   Match words on the left to related words on the right.

prevent	mandatory	remove		caution	protection	danger
warning	precaution	assess		take off	stop	evaluate
hazard	provide	prohibit		forbid	required	give

3   Discuss the differences in meaning with a partner.

Reading

1   Each of the posters *a–d* opposite has a different purpose. Look at them and complete these sentences.

   a   Poster ____ gives information about the correct way to do something.

   b   Poster ____ explains the responsibilities of the company and the workers.

   c   Poster ____ gives simple instructions in case of an accident.

   d   Poster ____ describes different types of safety warnings.

2   Look at the posters more carefully and decide whether these statements are true (T) or false (F).

   a   i   ____ Employers must train people in how to use safety equipment.
       ii ____ PPE needs to be specialised for some tasks.
       iii ____ Employers are responsible for putting away safety equipment.

   b   i   ____ An ambulance should be called in every case of a burn or scald.
       ii ____ You should not try to take off any of the injured person's clothes.
       iii ____ You should take off the person's watch.

   c   i   ____ A sign that tells you where fire equipment is kept is blue.
       ii ____ A yellow sign means some kind of danger.
       iii ____ First aid equipment will be found near a green sign.

   d   i   ____ There are five stages to lifting correctly.
       ii ____ If you can, you should use a machine to lift things.
       iii ____ You should look for dangers in your way before you lift.

3 **Find an example of each of the following verbs in the posters. Which nouns do they occur with?**

provide   consult   report   maintain   prohibit   assess

a

WHAT YOU SHOULD KNOW

PERSONAL PROTECTIVE EQUIPMENT

In situations where risks cannot be controlled by other means such as systems of work or engineering controls, employers are required to protect their employees from risks to health and safety by providing suitable personal protective equipment (PPE).

THE EMPLOYER MUST:
- Provide suitable PPE free of charge.
- Maintain PPE in working order and good condition.
- Provide relevant training in the use of PPE.
- Consult employees on suitability of PPE.

PPE PROVIDED MUST:
- Be relevant for the work undertaken.
- Protect effectively against particular risks involved.
- Fit properly and comfortably (adjusting in size where necessary).
- Not hinder the performance of any task.
- Not add to the risks involved.

THE EMPLOYEE MUST:
- Use the PPE provided.
- Report any loss, defects or damage to PPE.
- Take care to correctly store PPE when not in use.

b

WHAT YOU SHOULD KNOW

BURNS AND SCALDS

1. Place the burnt area under cold running water immediately for at least 10 minutes. If it is a serious burn ensure an ambulance is called.
2. If possible remove any items that may prevent swelling to burnt areas i.e. Belt, Boots, Watches or Rings.
3. Place a clean, sterile dressing over the burnt area.
4. Check that if required an ambulance has been called and check that the accident has been reported to the correct individuals.

1. DO NOT apply any lotion, ointments or creams.
2. DO NOT attempt to remove any items of clothing that may be sticking to the burnt area.
3. DO NOT touch or place anything other than a sterile dressing on a burn.
4. DO NOT burst any blisters that may form on or around the wound.

YOUR PROMPT ACTION CAN PREVENT SERIOUS INJURY OR EVEN DEATH!

EMERGENCY INFORMATION

HOSPITAL TEL: _____

DOCTOR TEL: _____

NEAREST FIRST AID: _____

YOUR FIRST AIDER IS: _____

c

WHAT YOU SHOULD KNOW

SAFETY SIGNS & THEIR MEANINGS

Recent regulation changes place a responsibility on employers to provide and maintain sufficient safety signs to warn of circumstances where risks to health & safety exist and to advise of precautions that need to be taken.

PROHIBITION SIGNS (DO NOT DO)
- A sign prohibiting behaviour likely to increase or cause danger eg. No Smoking. – Colour red.

MANDATORY SIGNS (MUST DO)
- A sign prescribing specific behaviour eg. Hard hats must be worn. – Colour blue.

SAFE CONDITION SIGNS (THE SAFE WAY)
- A sign indicating emergency exits or first aid/rescue equipment. – Colour green.

WARNING SIGNS (CAUTION, BEWARE)
- A sign giving warning of a hazard or danger. – Colour yellow.

FIRE SIGN (FIRE EQUIPMENT)
- A sign indicating the location of fire fighting equipment. – Colour red.

INFORMATION SIGN (GENERAL INFORMATION)
- A sign providing general information not covered by the above catagories.

d

WHAT YOU SHOULD KNOW

MANUAL HANDLING REGULATIONS

On 1st January 1993 regulations made under the Health and Safety at Work Act 1974 placed on employers additional responsibilities with regard to manual handling and lifting.

ASSESS THE SITUATION
- Can manual handling be avoided? If so use mechanical aid to lift or move the load.
- If not plan your lift.

CHECK FOR THE FOLLOWING
- Potential hazards ie. packing hooks, sharp edges, etc.
- Check stability of parcel, ease of grip, weight of parcel.
- Know your destination BEFORE lifting.
- Can you stop and rest if needed?
- Are you wearing protective equipment ie, gloves, safety boots, hard hat?

YOUR GUIDE TO SAFE LIFTING

1. Plan all aspects of your lift before taking weight.

2. Spread the feet, place them close to the load, bend the knees keeping back straight.

3. Grip the load firmly, keep arms close to the body, use the legs to lift upper body and load.

4. Hold load close to body, do not twist the trunk, move feet.

Speaking

Work with a partner. Discuss the following.

1 In Lesson 1 you saw what the colours of signs indicate. Do you remember them?

2 Look at the ten signs below. Which signs indicate:

   a that something is mandatory (must be done)?

   b that something is prohibited (must not be done)?

   c safety measures and first aid?

   d a warning?

3 Discuss what the signs mean and where you might see them.

Listening

1 🎧 Listen to six short conversations and number the six signs which are mentioned.

2 🎧 Now listen to some sentences from the recording. Try to write exactly what was said in that phrase or sentence.

   a make that mistake _____

   b another accident _____

   c I was underneath _____

   d sources of ignition _____

   e no protection _____

   f more windows _____

Pronunciation and speaking

1 **Look at the Skills Box. Conversation 1 begins like this. The stressed words are marked.**

  a Do you know where <u>Jack</u> is? I've been <u>waiting</u> for him for <u>twenty minutes</u>.

  b <u>Last</u> time I saw him, he was going off to wash his <u>hands</u>. He's been doing some <u>painting</u>.

Work with a partner and practise these two lines until the difference between the stressed and unstressed words is clear in your speech.

2 **Look at the tapescript for this lesson. Choose one of the other conversations and mark the stressed words. Then practise the conversation as before, concentrating on the difference between stressed and unstressed words.**

Writing

1 **Look at the four signs that were not discussed in Listening Exercise 2 above. Write a short dialogue between two or three people in the workplace which is connected with one of them. Start the first dialogue like this:**

A: What are you <u>doing</u>?

B: I'm going to …

A: Be careful …

2 **Practise your dialogue, focusing on which words in the sentences should be stressed.**

3 **Read your dialogue to some colleagues. They will tell you which sign goes with the situation.**

⇨Workbook pages 101–103

Lesson 3
Put out the fire

Speaking and reading

1 Discuss with a partner.

    a How can fires start? Think of as many causes of fire as you can.

    b What can you do to extinguish a fire?

    c What substances (apart from water) can be used to extinguish it?

2 What different types of fire are there? Match the classes of fire *A–D* with the example materials *i-iv*.

    Class A: fuelled by non-metallic solid materials

    Class B: involve flammable liquids and gases

    Class C: involve energised electrical wiring or equipment

    Class D: involve the combustion of unusual metals

    i   magnesium, sodium, titanium

    ii paper, wood, cloth, rubber, certain plastics

    iii gasoline, paint thinner, kitchen grease, propane, acetylene

    iv motors, computers, panel boxes, wiring, cabling

3 Work in groups of three.

    a Look at the table below. Check that you understand the column headings.

extinguisher		type of fire					
colour	type	solids (wood, paper, cloth, etc.)	flammable liquids	flammable gases	electrical equipment	cooking oils & fats	notes
	water	✓	✗	✗	✗	✗	*cools burning material reduces O_2*

    b Each member of your group should read one of the paragraphs opposite: Foam, Dry powder or CO_2. When you are ready, explain your text to each other.

c Listen to your colleagues' explanations and put a tick (✓) for *yes* and a cross (✗) for *no* in the table. Complete any extra notes or points of interest in the last column. The information for the water extinguisher has been done as an example.

Common types of fire extinguisher

Water Only suitable for Class A fires. Colour-coded red; work by cooling burning material as well as reducing oxygen.

Foam More versatile than water extinguishers. Can be used for Classes A & B fires
5 but not recommended for other classes of fire. Colour-coded cream. Work by forming a blanket or film on the surface. Because foam spreads over a wide area very quickly and is difficult to clean up, these are not recommended for use in the home.

Dry powder Multi-purpose fire extinguishers which can be used on Classes A, B & C fires (up to 1,000 volts maximum). Colour-coded blue. Work by "knocking down"
10 flames and depositing a layer of powder on the material.

CO₂ Ideal for Class C fires because CO_2 does not conduct electricity and does not support combustion. Can also be used on Class B fires, but unsuitable for Class A fires because they do not have a cooling effect. Colour-coded black. Work by preventing oxygen coming into contact with the material. Where there is a risk of a
15 particular type of Class D fire, a specific type of extinguisher should be available.

4 Work with a partner.

Student A: Describe a type of fire, e.g., engine oil on the floor.

Student B: Say what fire extinguisher is required to put it out (*foam, powder or CO₂*).

Vocabulary

1 Read the texts above about common types of fire extinguisher again.

a Underline the words *for* and *by* each time they are used.

b Circle the words which are used with *for* and *by* to form a fixed expression.

For example:

Only (suitable) for Class A fires.

2 Complete these sentences with *for* or *by*.

a A foam extinguisher works _____ covering the surface of the material.

b Water is unsuitable _____ Class C fires because it conducts electricity.

c CO₂ is recommended _____ Class C fires.

d The most suitable extinguisher _____ a fire which involves paper is water, but it cannot be used _____ Class B fires.

➪ **Workbook pages 104/105**

Lesson 4
Safety procedures

Speaking

All commercial aircraft are required to carry out certain procedures to ensure the safety of the passengers and crew. They often use specific equipment to do this.

1   **Make a list of procedures and the equipment they involve. For example:**

dealing with a sick passenger: a first aid box

2   **Compare your list with your partner's. What are the details of the procedures you have listed? How is each piece of equipment used?**

Listening

Now listen to a safety inspector talking about his job on a radio programme.

1   **Listen once and mark all the equipment and procedures on your lists which are mentioned. Which others are mentioned?**

2   **Complete the notes on the inspection procedure for each of the pieces of equipment.**

signs	1 make sure they are there 2
fire extinguishers	1 2 in right places 3
oxygen masks/bottles	1 2 in case of smoke in cabin
torches	1 2 batteries must be fully charged
sick bags	1 in case of turbulence 2
life jackets	1 2 in good condition
megaphone	1 2 crew to give instructions in an evacuation
seats	adjust up and down
seat belts	1 2 not twisted or frayed

Grammar

have to/need to

1 Look at the Language Box. How many examples of *have to*, *need to* and *be allowed to* can you find in the tapescript? Circle them.

2 Complete the following sentences with the correct form of *have to*, *need to* or *be + allowed to*.

a There are dozens of pieces of equipment that the inspector _____ check.

b Passengers _____ smoke on board.

c In case of emergency, crew members _____ give instructions to passengers.

d If an aircraft does not pass a safety inspection, it will _____ fly.

e The inspector _____ tell people all the details of his work.

f If there is smoke in the cabin, passengers will _____ use breathing equipment.

g The crew also _____ have safe seats and seat belts.

h All civil aircraft _____ carry a suitable first aid box.

> **Language Box**
>
> *have to/need to/be allowed to*
>
> When people talk about regulations and responsibilities in English, they often use the expressions *need to*, *have to* and *be allowed to*.
>
> Examples:
> Passengers *are not allowed to* walk near the aircraft on the ground.
>
> The inspector *has to* carry out detailed checks.
>
> Most of the safety inspection *needs to* be done while the cabin is empty.

Speaking

With a partner, describe as much as you can remember of the inspector's duties, using *have to/need to* and the verbs *check* and *make sure that*.

Example:

He has to check the seat belts.
He needs to make sure that the breathing equipment works properly.

➡️**Workbook pages 105/106**

Fasten seat belt

Reading

1   Look quickly at the text opposite. Which of these three titles is the best?

Types and functions of seat belts

Maintenance of restraints

Common restraint damage

2   Match the types of damage on the left with the correct definition on the right.

a	cut	___	**i**	general loss of shine, damage to surface
b	tear	___	**ii**	becoming longer
c	chafing	___	**iii**	marks caused by other substances
d	fraying	___	**iv**	the threads at the edge of the material are loose
e	stains	___	**v**	bending, twisting, change of shape
f	fading	___	**vi**	rust or oxidation
g	distortion	___	**vii**	damage with a sharp tool, e.g., a knife
h	elongation	___	**viii**	loss of colour over time
i	cracking	___	**ix**	general damage caused by use over a long time
j	corrosion	___	**x**	damage caused by rubbing against other objects
k	wear	___	**xi**	long, narrow breaks or splits
l	deterioration of finish	___	**xii**	damage caused by pulling, e.g., paper, in two directions

Speaking and writing

1   Look at the pictures opposite. With a partner, discuss what kind of damage is shown.

2   Look at the damage report below. Which belt does it describe?

item	notes
belt label	OK
damage to belt straps: specify place and type	signs of fraying around securing bolts
damage to fixings/fastenings: specify place and type	none
belt adjustment	difficult to tighten and loosen
damaged item: Repair/Discard	discard

3   Write similar reports for the other belts.

Seat belts are designed to withstand extremely high loads, e.g., a 100-kg man travelling at several hundred kilometres an hour. But if there is any damage to a belt, its load-bearing ability will be significantly reduced. Therefore, as part of regular maintenance, all of the restraints on an aircraft must be checked in accordance with a

5 set procedure, such as the following:

1 Check that the seat belt label is intact and legible.

2 Without removing the safety harnesses from the aircraft, thoroughly scrutinise each individual strap of the safety harnesses in both front and rear cockpits for any evidence of:

10   a breaks in stitching

    b cuts and tears

    c chafing

    d fraying

    e stains due to acid, oil, grease and water

15   f colour fading due to exposure to sunlight

    g distortion of end fittings

    h signs of excessive wear or elongation of the attachment holes

3 Examine each attachment bracket, its securing bolts, saddle washers and the fuselage frame in the vicinity for evidence of cracking, corrosion, wear or

20 deterioration of the surface finish.

4 Finally, test all restraint mechanisms to ensure that the restraint can be:

    a fastened

    b released

    c tightened

25   d loosened

5 In the event of the restraint failing any of the above checks, it must be replaced and either discarded or sent to an authorised repairer.

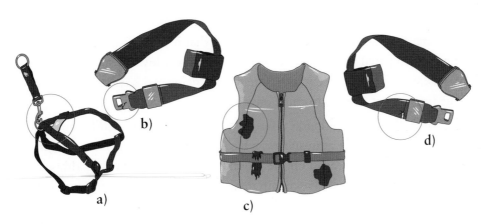

⇨**Workbook pages 107/108**

Lesson 6
Fuel + heat + O₂ = fire

Speaking

1 With a partner, discuss what is shown in this diagram.

2 What happens if each of the three main elements in turn is removed? How can this be done in real situations?

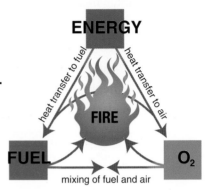

Reading and writing

1 Read the description below of the firefighting gas halon. Which other gas or gases is halon compared with?

Halon differs from all other **extinguishing** agents in the way it puts out fire. Its essential extinguishing ability lies in its **capacity** to chemically react with the oxygen and put out the fire immediately, without leaving the kind of mess and damage that can be caused to avionics and electrical equipment by water, foam or dry powder. Halon fire

5 extinguishers, which are effective on the three most common classes of fire, A, B and C, are thus ideally suited to aircraft use. Although they are more expensive than other types, the results are well worth the difference in price.

Like carbon dioxide (CO_2), halon is **unsuitable** for use in open areas. If you spray it into the open air, it disperses almost as soon as it is sprayed, but it is highly effective in

10 closed areas. However, halon is superior to CO_2 for three good reasons. CO_2 must be stored at high pressure, whereas halon is stored at low pressure. This means that halon fire extinguishers can be less than half the size and weight of equivalent CO_2 extinguishers, which makes them particularly suitable for use aboard aircraft, where space and weight savings are a priority. CO_2 works by physically displacing all of the

15 oxygen in the compartment, which suffocates the fire. Unfortunately, because it is heavier than air, it can also suffocate people. Halon, on the other hand, works by chemically **interrupting** the burning process, which means that it requires a fraction of the amount to do the job compared with CO_2 and there is therefore a lower risk of **suffocation**. In areas where there are people present, the systems are engineered to provide sufficient halon to

20 kill the fire but not enough to cause anyone harm.

Unfortunately, like CO_2, halon is not good for the environment because it damages the ozone layer of the upper **atmosphere**, which protects the Earth from **dangerous radiation**. Its use is forbidden on the ground in most countries and, although it is still used in many **military** and civilian aircraft, it is gradually being replaced by new gases

25 which are more environmentally friendly.

2 Look at the list of characteristics below. Refer back to the comparisons in the text and mark each characteristic **H, CO₂, both** or *neither* as appropriate.

a can be used in open areas *neither*

b stored at high pressure _____

c small extinguisher _____

d heavy extinguisher _____

e no residue which can damage machinery _____

f prevents oxygen coming into contact with fuel _____

g changes chemistry of oxygen _____

h small amount required to put out fire _____

i higher risk of harm to humans _____

j not environmentally friendly _____

3 Complete the table with your own notes for the two types of fire extinguishers.

	CO_2	halon
class of fire	B, C	
how it works		
open areas	unsuitable	
closed areas		
storage pressure		
weight of extinguisher		$< ½ CO_2$
size of extinguisher		
danger to humans		
amount of gas required		much less than CO_2
environment		
cost	?	more expensive than others

Pronunciation

Look at the words in bold in the text. Put them in the correct column of the table according to their stress.

Ooo	oOoo	ooOo

➡ **Workbook pages 109/110**

You are going to read information from an advertisement for an automatic fire prevention system for military aircraft.

Reading and vocabulary

1 **Guess the missing words from the nine expressions below. Check your answers in the text opposite.**

a o _ _ -rich fuel vapours

b a _ _ filter

c m _ _ _ t _ _ _ _ c _ and life cycle costs

d a continuous f _ _ _ of nitrogen

e f _ _ _ tank ullage

f c _ _ _ member safety

g vulnerability to enemy ground f _ _ _

h combustible fuel/air m _ _ _ _ _ _

i sloshing of fuel in the t _ _ _ _

2 **Discuss the possible meanings of any unfamiliar words in the expressions.**

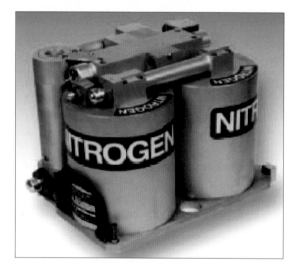

3 **Match the following titles to the sections of the advertisement A–E.**

Specifications   Adaptability   Solution   Features   Risk

4 **Which of these best summarises the information in the advertisement?**

a Ullage in helicopters is very volatile due to lightning and static discharge. OBIGGS removes nitrogen from the fuel tanks, so that the vapor is inert against all ballistic threats. Maintenance of the equipment is carried out by ground crews.

b The vapour in the ullage is extremely volatile. To avoid explosion, the OBIGGS pumps an inert gas, nitrogen, into the tanks to replace the missing fuel.

c OBIGGS burns oxygen-rich fuel vapors by mixing them with liquid nitrogen. This meets helicopter production and operations requirements. Maintenance costs are low, but the equipment is very large and heavy and tends to vibrate.

NC1147 On-Board Inert Gas Generating System (OBIGGS)

A As the amount of fuel in the fuel tanks decreases, the empty space (ullage) in the tanks increases. Combustion of oxygen-rich fuel vapors in the fuel tank ullage can cause explosions and fires. The risk is increased in helicopters due to sloshing of
5   fuel in the tanks, fuel tank vibration and vulnerability to enemy ground fire.

B OBIGGS increases aircraft survivability and crew member safety by inerting the aircraft fuel system against ballistic threats up to and including 23 mm HEI (High Explosive Incendiary) rounds and environmental threats such as lightning and static discharge. OBIGGS inerts the aircraft fuel system by providing a continuous
10   flow of nitrogen-enriched gas into the space in the fuel tanks, thus eliminating the combustible fuel/air mixture.

C This OBIGGS is being produced for the MH-47E Chinook and is designed to accommodate your specific attack, utility, cargo and reconnaissance helicopter inerting and fuel system ballistic tolerance requirements.

15 **D** _ Built-in test capability

   _ Integral inlet air filter, system controller and oxygen monitor

   _ Quality product gas is produced immediately on start-up

   _ Requires no ground support equipment

   _ Compared with foam, assures increased usable fuel and easy access to fuel
20   tanks by ground crews

   _ Maintenance and life cycle costs are both significantly lower than other inerting alternatives

E Air consumption: 2.2 lbs/min. @ 40 psi

   Electrical: 28 V DC – 25 W

25   Dimensions: 10.18 x 8.63 x 12.93 in. (26 x 22 x 33 cm)

   Weight: 30.0 lbs (13.6 kg)

http://www.carletonis.cm/productsbu/nc1147.htm

Writing

Summarise the information in each section of the advertisement in one brief bullet point. For example:

- Section A: Ullage vapours volatile, esp. in helicopters
- _____
- _____
- _____
- _____

⇨**Workbook pages 111/112**

You are going to listen to an ex-US Navy pilot talking about oxygen equipment for pilots flying at high altitude.

Vocabulary

1 Find two words on the right with a similar meaning to the ones on the left. Copy them next to the words on the left.

a hose (*n*) _____

b internal (*adj*) _____

c external (*adj*) _____

d bent (*adj*) _____

e tough (*adj*) _____

f flexible (*adj*) _____

g block (*v*) _____

> inner   hard   outer   soft
> exterior   crimped   rigid
> pliable   interior   stop   close
> off   tube   folded   line

2 Discuss with a partner. Do all three words have exactly the same meaning in each case? What are the differences?

Listening

1 🎧 Listen once to the recording and answer these questions.

a Which system was causing problems in this case, LOX or OBOGS?

b Was it a design fault, a maintenance problem or pilot error?

c Was it just one aircraft or a general problem?

2 🎧 **Listen again. Choose the correct information from the two options in each case.**

a ____ With the LOX system, oxygen is stored on board the aircraft.
b ____ With the OBOGS system, oxygen is stored on board the aircraft.

c ____ LOX is more reliable than OBOGS.
d ____ OBOGS is more reliable than LOX.

e ____ The OBOGS system worked correctly when the planes were on the ship.
f ____ The OBOGS system worked correctly at high altitude.

g ____ The mechanics were following the procedure in the manual.
h ____ The mechanics thought the OBOGS hose was the same as the old LOX hose.

i ____ The warning light was connected to an oxygen monitor downstream.
j ____ The warning light went on if the oxygen level inside the concentrator was low.

3 **Compare and discuss your answers.**

4 **Complete the paragraph below, using vocabulary from the box.**

> flexible   blocked   crimped
> internal hose   rigid   intact   bent
> external   hose   sleeve   lines   damage

The concentrator on the OBOGS oxygen supply was tested and had no defects. However, when the _____ on one of the aircraft was inspected, the metal-braided _____ was broken, and the true condition of the _____ could easily be seen. It was found to be crimped, which meant that the flow of oxygen was partly _____. Several of the other aircraft were also found to have _____ or _____ oxygen lines. The reason was that the mechanics who maintained the system thought that the OBOGS line was _____, like the old LOX lines. In fact, OBOGS lines are _____. To remove the concentrator, they bent the inner plastic _____ forcefully out of the way. This action crimped the hose, although the _____ was not seen because the _____ sleeve remained _____.

Speaking and writing

Write three or four questions to ask your partner about the problem. Take turns to ask and answer questions using your own words. Use the diagrams below to help you.

rigid pipe

flexible pipe

➡️ Workbook page 113

Lesson 9
Split-second action

Vocabulary

1   Work with a partner. Look at the series of illustrations below and, using your own words, describe what it shows.

2   Label the following in the drawings.

> main parachute   seat separates   rocket motor   drogue parachute   restraints

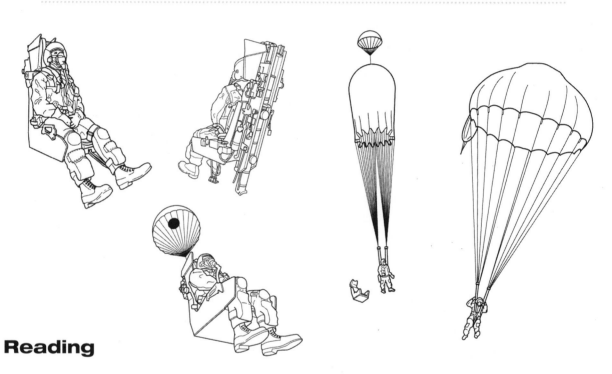

Reading

1   Look at the following notes, which describe the sequence. Then read the text on the opposite page to put them in the right order.

> ____ Normal parachute descent.
>
> ____ Seat rises with initial thrust: limb restraint cords operate.
>
> ____ Rocket motor ignites.
>
> ____ Firing handle pulled to initiate escape.
>
> ____ Main parachute deploys, automatic seat separation.
>
> ____ Drogue parachute stabilises seat at 1.0 second.
>
> ____ Rocket motor burns out at 0.45 seconds, drogue gun piston fires at 0.5 seconds.

Emergency ejection

Once the decision has been made by the crew to leave the aircraft, the crew will activate the ejection seat by pulling the seat-firing handle connected directly to a unit under the seat pan. The ejection seat is mounted on the combined guide rails and telescopic
5 ejection gun unit, which in turn is attached to the aircraft structure. The ejection gun provides the initial thrust for seat ejection, propelling pilot and navigator in their seats from 0–160 m/h, in a quarter of a second. While being propelled through the air by the ejection seat, an average 65-kg man would equal a 2,000-kg weight. This is equivalent to 30 g, and is only made possible by flying with special safety clothing, and the
10 reinforced straps and restraints holding the pilot and navigator firmly to their seats.

A delay mechanism provides a time delay between the ejection of each seat, and this ensures that the pilot and navigator do not collide during ejection. The thrust of the ejection is sustained by a rocket motor under the seat pan as the seat leaves the ejection gun. This ensures the seat reaches a sufficient height to enable the parachute to deploy
15 and open out, even if ejection is initiated at zero speed at ground level. At ejection plus half a second, the drogue parachute fires: the drogue provides stability and also pulls the main parachute out of its housing. Once the seat is stable, the main parachute opens and the seat is automatically discarded. From this point on, the descent is the same as a normal parachute descent. If ejection is over water, crews have to release
20 themselves from the parachute harness and climb into a self-inflating dinghy.

2 What do these numbers and quantities refer to?

160 m/h   30 g   0 m/h   0.5 sec.   65 kg   0.25 sec.   2,000 kg

Vocabulary and speaking

1 Find the following verbs in the text and look at how they are used to express purpose.

provide   sustain   ensure   enable

2 With a partner, ask and answer questions about the purpose of the different elements in the ejection system, for example:

Student A: What does the drogue parachute do?

Student B: It provides stability.

⇨Workbook pages 114/115

Lesson 10
Plan B

Vocabulary and speaking

1   If engine power fails, which of the systems listed in the box are important for aircraft and pilot survival? Why? Discuss your ideas with a partner. Try to use some of the expressions below.

> radio   landing gear   high-altitude oxygen   flaps   pilot suit heating
> weapons systems   cameras   air brakes   rudder and ailerons   backup systems

2   With your partner, decide on the two most important items in the list above.

3   Replace the bold words in the text with words and phrases from the box.

> basic jobs   secondary   usually   act as replacements
> unplanned   makes it possible for   fail

> Modern aircraft are usually fitted with **backup** systems called emergency packages, which can **take over** when main systems **go down**. For example, an aircraft's hydraulic system is **normally** powered by the engine, but if this has stopped for some reason, an emergency package can perform **essential functions** such as extending the flaps and lowering the landing gear. This **enables** the pilot to make an **emergency** landing without engine power.

4   Were your ideas from Exercise 1 mentioned?

Reading

1   Quickly read the text opposite. Find and underline the words and phrases in the box.

> accumulator
> oil head
> charging value
> hydraulic lines to the landing gear and flaps
> pipeline from the system pressure line

Emergency package: flaps/landing gear

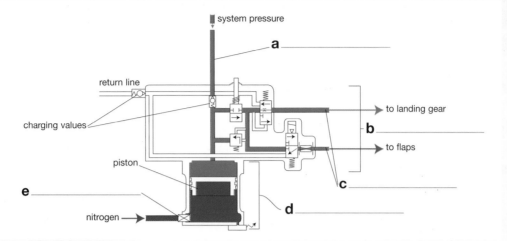

- The emergency package comprises an **accumulator** and an **oil head**. The body of the accumulator is in the form of a cylinder, which contains a cup-shaped piston.

5
- The accumulator is filled by introducing nitrogen through a **charging valve**. A pressure transducer transmits information about the nitrogen pressure in the accumulator to the secondary instrument display panels in the cockpit.

- The other side of the piston forms an oil head containing restrictors, check valves, pressure relief and pressure release valves, plus emergency valves serving **hydraulic lines** to the landing gear and flaps.

10
- When the main engine-powered hydraulic system is operating, a pipeline from the **system pressure line** directs hydraulic pressure fluid to the emergency package. This maintains the nitrogen pressure at 3,000 psi.

- If the main hydraulic system goes down for any reason, sufficient hydraulic fluid is retained in the oil head to open the main landing gear doors, lower the landing
15
gear and extend the flaps. The piston, driven by the high-pressure nitrogen in the accumulator, pressurises the fluid at 3,000 psi, enabling it to operate the flap and landing gear actuators.

2 Read the text again and add the underlined items as labels to the drawing.

Speaking

With a partner, discuss what sort of backup system would be used if the following systems failed.

- the air brakes
- the radio
- the high-altitude oxygen system
- the rudder and ailerons

➡ Workbook pages 115–118

Looking for damage

It is very important that maintenance crews check tyres for correct pressure and for damage.

Speaking

Discuss with a partner.

    a Why is it important for a tyre to be as soft as possible?

    b Why is it important for it to be as hard as possible?

    c Why is it important to check for damage to a tyre?

Vocabulary

1   **Put the words in the box into the correct column in the table.**

> tread   misalignment   bead   sidewalls   burst
> damage   vibration   cracking   layers   wear   puncture

parts of a tyre	problems with tyres

2   **Mark the stress on the multi-syllable words and practise saying them with a partner.**

Listening

1   🎧 Read the job sheet opposite. Listen and complete the **JOBS CARRIED OUT** section.

2   🎧 Listen again and complete the **REMARKS** section with any relevant comments or information you hear.

JOB SHEET	TYRE INSPECTION	
REASON FOR INSPECTION	BAD LANDING SUSPECTED DAMAGE TO TREAD	
		WRITE YES OR NO
JOBS CARRIED OUT	TREAD DEPTH MEASURED	
	TYRE DEFLATED	
	FOREIGN BODY REMOVED	
	WHEEL REPLACED	
	WHEEL FLIPPED	
	BEADS CHECKED	
	SIDEWALLS INSPECTED	
REMARKS AND FURTHER WORK		

3 🎧 From the conversation, do you remember what *it* is in each of these phrases? Write down your ideas, then listen again and check your answers.

a put it in the grooves

b look at it

c demount it and turn it round

d remove it

e inspect it

Language

1 Look at the Language Box. Complete these sentences from the conversation using *some/any*, *something/anything*.

a We need to look for _____ stuck in the tyre.

b Can you see _____? _____ foreign matter?

c What happens if we find _____ stuck in the tyre? Do we try to remove it?

d That's a bad sign. It means that there's _____ damage.

2 Read the tapescript to check your answers.

➡ Workbook pages 119/120

Language Box

some/any

Any refers to the whole group of items which are available, e.g.:
You can contact me any time.
If you have any problem at all, talk to the supervisor.

Some refers to a certain part of the whole group, e.g.:
Can I take some of this paper?
Some people think it wasn't a good idea.

These words work in the same way:
something, somebody, somewhere;
anything, anybody, anywhere.

Lesson 2
Assemblies and systems

Another external application for pneumatics is found in the landing gear of many aircraft.

Vocabulary

1 **What is the function of a landing gear assembly? Discuss your ideas with a partner, using some of the words in the box below.**

dampen   vibration   impact   brake

2 **Look at figures 1 and 2, which are views of the same assembly. Discuss how you think the system works.**

3 **Match the halves of these sentences and see if you were right.**

a The landing gear assembly consists of

b One function of the assembly is to absorb

c This system uses nitrogen gas under pressure because

d In addition, the lines contain pressurised hydraulic

e A piston reciprocates

i inside a cylinder barrel.

ii fluid which moves actuators.

iii it is much lighter than a metal spring.

iv the landing gear leg and the wheel and tyre assembly.

v the shock of an impact or provide damping.

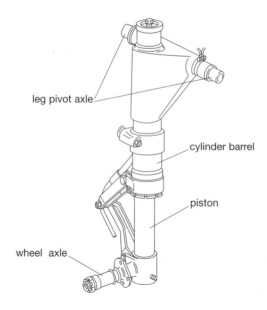

Figure 1

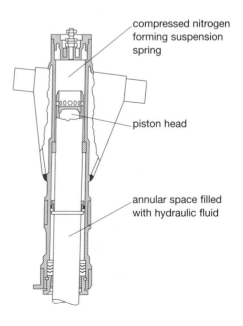

Figure 2

Reading

1 Read the texts below quickly. Which one explains the system better?

Text 1

The leg is an air and oil shock strut assembly, consisting of a cylinder and piston assembly. As well as having a cylinder of compressed nitrogen to absorb the shock on landing, or the demands of moving along a non-level surface, the assembly contains hydraulic fluid in the space around the piston to ensure damping during the operation of the piston so that the strut does not recoil too quickly, which would cause vibration. This illustration shows a landing gear leg from a small aircraft.

Text 2

- The <u>drawing</u> is a shock strut assembly.
- The assembly uses a combination of oil and gas to absorb landing <u>impact</u>.
- The space above the piston head is filled with nitrogen gas <u>under pressure</u>.
- The <u>annular space</u> between the piston body and the <u>barrel</u> is filled with hydraulic fluid.
- The nitrogen provides the spring to absorb both landing and taxiing loads, particularly on <u>uneven</u> surfaces.
- The hydraulic fluid, which is displaced through flow restrictors during piston <u>reciprocation</u>, provides the required damping during both the compression and extension phases of the strut.
- This avoids vibration caused by excessively rapid <u>return</u> of the <u>strut</u>.

2 Read the texts again. Circle a word or phrase in the first text with the same meaning as the underlined words in the second, e.g., <u>drawing</u> – *illustration.*

Vocabulary

1 How many expressions can you find in the text with *of*, e.g., *a combination of.*

2 Complete the sentences with the expressions in the box.

> instead of   the demands of   in case of   a combination of

a Most modern aircraft are fitted with emergency packages _____ systems failure.

b For ignition, _____ fuel, heat and oxygen is required.

c Nitrogen is used _____ oxygen because it is inert, and so safer.

d The Hawk responds well to _____ acrobatic flight.

⇨Workbook pages 120–122

Lesson 3
Up, up and away

Unlike liquids, gases can be compressed into a smaller volume, and if a quantity of gas can be compressed, it can also be expanded.

Vocabulary

1   Match the verbs on the left with the definitions on the right.

a	occupy	**i**	get physically smaller; shrink
b	increase	**ii**	rise in amount
c	decrease	**iii**	get physically bigger; grow
d	expand	**iv**	use a space
e	contract	**v**	go down (in quantity)

2   a   Which of these can increase and decrease, and which can expand and contract?

> temperature   volume   a balloon   pressure
> cost   height   a tyre   halon   wood

b   What do these units refer to?:

> psi   Pa   °K   °C   °F   cm³   cubic in.   m³

c   How do you say each one?

Speaking and listening

You are going to listen to a conversation between the owner of a hot air balloon and his two passengers.

1   Look at the picture of a balloon. In pairs, discuss what each labelled part is for.

2   🎧 Listen to the conversation and tick the information mentioned about each part.

the burner
_____ it burns nitrogen
_____ it burns propane
_____ it is always a double unit

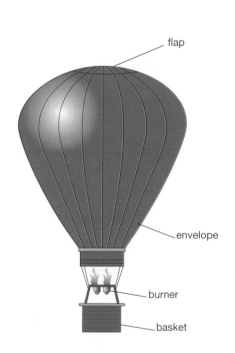

flap

envelope

burner

basket

the envelope

___ it is filled with hot air

___ it measures about 10 x 20 m

___ it is blue

the flap

___ it is at the top of the balloon

___ it is connected to the basket by a rope

___ it lets in warm air

the basket

___ it can carry four passengers

___ it is tied to the ground

___ it is made of fabric

3 Read these three paragraphs about how the balloon works. Work with a partner, but do not write anything yet. How much can you complete between you?

A Charles' Law

Charles' Law states simply that the _____ of a gas is directly _____ to its temperature. This means that if you increase the _____ of a gas, its volume will increase.

B Why hot air rises

According to Charles' Law, a gas occupies more space when it is _____ than when it is _____. As it expands, it becomes less _____, i.e., lighter. When the air in a balloon is hot, it is less dense than the air outside the balloon. It becomes light, and the balloon _____ .

C Descending

If you stop using the burners, the air in the balloon _____ . As it cools, it _____ (according to Charles' Law) and gets heavier: the balloon _____ . If you open _____ at the top of the balloon, hot air _____ and the process of cooling is speeded up.

4 🎧 Listen again to the recording and complete the missing information.

Reading and speaking

Look at the formula for the Combined Gas Law below and discuss the questions.

$$\frac{P1 \quad V1}{T1} = \frac{P2 \quad V2}{T2}$$

1 What do P, V and T stand for?

2 Discuss how you would say the formula in words.

⇨ Workbook pages 123/124

Vocabulary and speaking

Choose a verb from the box to complete each phrase.

arm	return	fill	purge
check	consult	leave	set
prime	correct	cock	use

a _____ the lever to its original position

b _____ only the amount of oil stated on p6

c _____ oil levels to ensure they are the same

d _____ page 62 of this manual

e _____ the valve open

f _____ the system of air

g _____ the reservoir with oil

h _____ any problems

i _____ the fuel valve

j _____ the dial to the zero position

k _____ the protection system

l _____ the pump

Reading

1 Look at the diagram of a compressor. What uses can you think of for compressed gas?

2 Read the instructions from the maintenance manual on the opposite page and try to identify as many parts as you can. NB: Some parts are not shown.

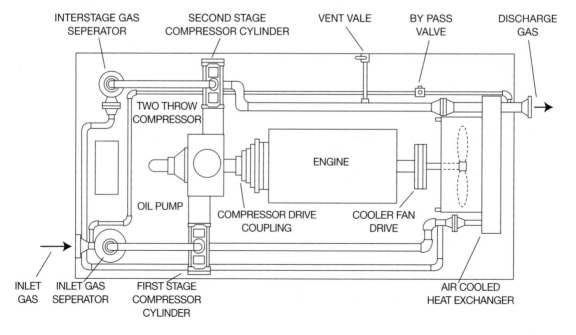

Prior to Starting

(See Figures 1 and 5)

1. Purge entire system as follows:
   (a) Open the suction block valve.
   (b) Open the bypass valve.
   (c) Open the vent valve.
   (d) Close, then reopen the bypass valve to ensure that all air has been removed from the system.
   (e) Finally, close the vent valve and leave the bypass open for start-up.
2. Check compressor oil level and engine oil and water levels.
3. For initial start-up:
   Prime the compressor oil pump by adding oil through the top 1/4" NTP hole on the oil pump end-housing. Use a standard oil can and the correct grade of oil. Fill the oil cooler and all the piping to and from it. For more information, see page 6, "Oil Requirements".
4. Check all shutdown tattle tales to be sure they are armed.
5. Cock the fuel valve by pushing the lever in the direction of gas flow.

6. Turn the start-run timer to 3–4 minutes, locking out the green-tagged tattle tales.

Starting

1. Set the throttle valve per the engine manufacturer's instructions. Turn the ignition switch to the run position and start the engine.

After Starting

1. Once the unit is running, return the start-run timer to the "0" position, this arms all sensors and protects the system.
2. Check oil pressures to be sure they are correct. See the "Oil Pump" section of this manual for oil pressure adjustment procedures.
3. Inspect the entire system for vibration, noises, leaks, and oil and coolant levels. Correct any problems.
4. After the engine has warmed up (water temperature of 165°F), begin loading the unit by opening the discharge block valve while slowly closing the bypass valve. Once the bypass valve is closed the compressor is on-line and delivering gas.

3 Read the instructions again carefully.

a How do you:
  – prime the compressor oil pump?
  – cock the fuel valve?
  – load the unit?

b Why do you need to:
  – close and reopen the bypass valve?
  – check all shutdown tattle tales?
  – check oil pressure?

c What is the importance of:
  – page 6 of the manual?
  – 165°F?
  – 3–4 minutes?

Language Box

before/after/once/while

Maintenance procedures must follow a sequence. This can be indicated by expressions *before*, *after*, *once* and *while*.

NB: Each expression is used with different grammatical forms:

Hold the assembly *while* turn*ing* the screw anticlockwise.
After the air *has* cool*ed*, open the valve.
Once the water *is* hot, switch on the unit.
Before connect*ing* the power supply, ensure the unit is switched off.

Language and speaking

Look at the Language Box. Then work with a partner to do Exercises 1 and 2.

1 In the text at the top of the page, circle all instances of these expressions of time.

> after   once   while   finally

2 Ask and answer questions with a partner beginning *When ...?*

Example: Student A: When do you open the vent valve?

   Student B: After you've opened the bypass valve.

➡**Workbook pages 124/125**

In the fridge

Vocabulary and speaking

There are three physical states: solid, liquid and gas.

1 Discuss with a partner: what change of physical state is described by these verbs? For example:

Contract means to get smaller. Many solids, e.g., metals, contract when they are cooled.

> evaporate   liquefy   freeze   solidify   vapourise   melt   condense
> give off heat   absorb heat   cool down   heat up   expand   contract

2 Which is the stressed syllable in the words above? Practise saying the words aloud.

3 With your partner, find examples of all of the processes in Exercise 1 from everyday life. For example, *You can store solid and liquid foods by freezing them.*

Listening and speaking

You are going to listen to a conversation between Ali and his technical instructor, Mr Bashir, after a science lesson on heat energy.

1 Look at the diagram of a refrigerator and discuss whether or not you agree with these statements. Mark them Y (yes), N (no) or DK (don't know).

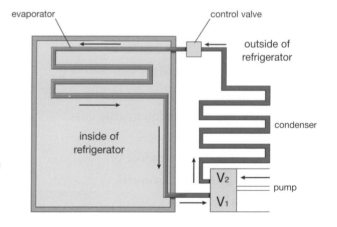

a The refrigerant is a special fluid inside the refrigerator pipes. ____

b In the evaporator, the refrigerant changes from vapour to liquid. ____

c The pump circulates the refrigerant. ____

d In the condenser, the refrigerant turns into liquid. ____

e The refrigerant absorbs heat in the condenser. ____

f Leaving the fridge door open will cool the room. ____

g On the compression stroke, valve V2 opens. ____

h The pressure in the condenser will be higher than in the evaporator. ____

i The control valve has a large hole (orifice) for liquid to pass through. ____

j A heat pump transfers heat from one place to another. ____

2 🎧 Listen to the conversation and listen for Mr Bashir's explanation of how the fridge works.

3 Talk to a partner. Are *a–j* above true or false?

Pronunciation

1 Look at the Skills Box. Then read the extract below from the recording aloud with a partner.

2 When you read out the text, where are the pauses? Mark them /.

3 🎧 Listen and check whether you were right.

4 Again, practise reading the text aloud, taking care to pause at the marks '/' but not in any other places.

Well OK, basically a heat pump uses the fact that when a liquid evaporates it needs to absorb heat from its surroundings and when a vapour condenses it gives off heat the best example is an ordinary domestic refrigerator the inside is cold but the grill at the back is warm the heat is removed from inside the refrigerator and then given off into the room which is why you can't cool down a room by leaving the refrigerator door open. Because when you leave the door open, the refrigerator works hard to try to cool the room and at the same time it's giving off the same heat back into the room heating the place up.

Speaking

Cover the text above. Look at the diagram. In pairs, take turns to ask and answer questions about how the system works.

➪Workbook pages 126–129

Lesson 6
On cloud nine

Vocabulary and speaking

1 Complete the missing letters in the words below. All the words are from previous lessons in this unit.

ev _ _ _ _ te   co _ _ _ _ se   ab _ _ _ b   ex _ _ nd   con _ _ _ ct   tem _ _ _ _ _ ure

2 Write the noun and verb forms to complete this table.

verb	noun
condense	
	absorption
humidify	
	weight
evaporate	
	content
measure	
	reading
moisten	

3 Work with a partner. Take turns to explain to each other the meaning of the words in the table. Do not write anything.

Reading

1 Look quickly at the text below and match each paragraph with one of these titles.

Measuring humidity   Atmospheric humidity   Relative humidity

Controlling humidity   Absolute humidity

The humidity – the moisture content – of a volume of air is usually denoted by the
mass of water vapour contained within it in grammes per cubic metre (g/m^3). When air
contains the maximum possible amount of moisture, it is said to be saturated, and
cannot absorb any more water vapour. *Absolute* humidity is the term used for this

5 amount of vapour. It is the temperature of a volume of air which determines how
much vapour it can hold: the hotter the air, the more moisture it can hold. Figure 1
shows the variation of absolute humidity with temperature.

However, most air is not usually saturated and could absorb more moisture. Relative

humidity is measured by comparing
10 the actual mass of vapour in the air
with the mass of vapour it could
contain at the same temperature. For
example, air at 10°C contains 9.4 g/m³
of water vapour when saturated. If air
15 at this temperature contains only 4.7
g/m³ of water vapour, then the relative
humidity is 50%.

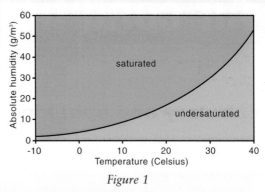

Figure 1

Because it is not practicable to weigh air, an instrument called a hygrometer is often
used to measure relative humidity (Figure 2). A hygrometer has two thermometers,
20 one dry bulb at standard air temperature and one wet bulb thermometer. The wet bulb
thermometer is an ordinary thermometer which has the bulb covered with a moist

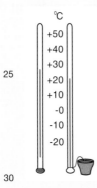

Figure 2

cotton bag. Evaporation of water from the muslin lowers the
temperature of the thermometer. The difference between the two
temperature readings of the hygrometer is the relative humidity, so if the
25 wet bulb thermometer reads 20°C and the dry thermometer reads 25°C,
the relative humidity of the air is 5/20, or 25%.

Some water in the form of invisible vapour is intermixed with the air
throughout the atmosphere. It is the condensation of this vapour which
causes clouds, rain, snow and fog and contributes greatly to our climate
30 and weather: excessive atmospheric moisture can cause very difficult
weather conditions both for operations on the ground and for air traffic.

2   **Read the text again and choose the correct definition for each of these terms
according to the text.**

a hygrometer:
– a device that measures air temperature by using a wet bulb
– a device for measuring relative humidity using a comparison
– a kind of thermometer which lowers the absolute humidity of the air

b absolute humidity:
– the water content of a volume of air at saturation point
– the amount of water in a cubic metre of air
– the hottest that air can get before it becomes saturated

c atmospheric humidity:
– the maximum amount of cloud, rain, fog and snow in the atmosphere
– the combination of water vapour and air in the atmosphere
– condensation of water vapour intermixed with air

d relative humidity:
– the amount of vapour, in g/m³, in a volume of air at 10°C when it is saturated
– half of the absolute humidity
– the mass of water in a volume of unsaturated air compared with its absolute humidity

⇨**Workbook pages 129/130**

Lesson 7

Air conditioning

Vocabulary and speaking

1 **Match the definitions with the parts of an air conditioning unit.**

a reheater	**i** A device or container that decreases the temperature of something.
b cooler	**ii** A machine or device for combining and mixing substances.
c hygrostat	**iii** An instrument or device that regulates temperature.
d thermostat	**iv** An instrument that measures and controls the relative humidity of the air.
e fan	**v** A device for separating and removing solid particles from liquids and gases.
f mixer	**vi** A device that heats a substance again after it has been cooled.
g filter	**vii** A device, usually with rotating blades, that circulates currents of air.

2 **Look at an air conditioning unit in the room where you are.**

a Are any of the components above visible?

b How do you think the system works?

Reading

1 **Look at the schematic below. Which of the components in Exercise 1 above might the boxes represent?**

2 **Read the text and complete the schematic by labelling the boxes 1–4.**

Air conditioning

Definition

Air conditioning (AC) is a method of providing a comfortable inside atmosphere by automatically controlling the temperature and humidity of the air and by recirculating,
5 freshening and cleaning it. The operation of most air conditioning systems can be represented by five main units, as shown in Figure 1 below.

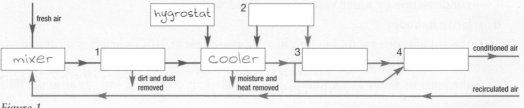

Figure 1

Operation

- First, the system draws in recirculated air and mixes it with fresh air.

- The freshened air passes through a filter which removes dust and dirt.

10 • Next, the air must be cooled to remove excess moisture so that the humidity is reduced. The air passes across a cooling coil, similar to the one in a refrigerator, which lowers the temperature. This causes excess moisture in the air to condense. Like a refrigerator, the AC has a compressor in order to keep the coils cooled.

15 • After cooling, the air may need to be reheated. In this case, it is passed through a hot-water coil or an electric heating element.

- Finally, the air is returned to the air-conditioned area by the fan. A thermostat and hygrostat are necessary because the outside temperature and humidity may change. These devices
20 measure the temperature and humidity, respectively, compare the measured values with the required values, then activate the
25 reheater or cooler if there is a difference.

- Fig. 2 shows a cutaway pictorial view of a typical simple AC window unit
30 used in hot countries.

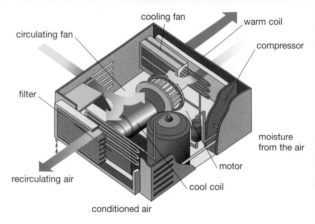

Figure 2

Writing and grammar

1 **Look back at the text and circle these expressions for describing reason and purpose.**

> which   this causes   so that   is a method of   in order to   to   because

2 **What type of words are they used with (nouns, verbs, ~*ing* forms, etc.)?**

3 **Write sentences answering these questions.**

a Why does the moisture in the air condense?

b Why is a hygrostat required?

c How is the correct temperature ensured?

4 **Work in pairs. Write three or four questions each about the AC system above and then take turns to ask and try to answer them.**

⇨ **Workbook pages 131/132**

Vocabulary and pronunciation

1 Work with a partner. One of you should check the meaning of the words in list A below; the other should check list B.

2 Take turns to explain the meaning of the words to each other. Use your own words as much as possible: do not memorise the dictionary definition.

A	B
replenish (*v*)	circulation (*n*)
percentage (*n*)	replace (*v*)
dilute (*v*)	filter (*n*)
contaminant (*n*)	recirculate (*v*)
ventilate (*v*)	interval (*n*)

3 Divide the words into three groups according to the number of syllables they contain (2, 3 or 4).

4 Practise saying the words with the correct stress and pronunciation.

Reading

The text on the opposite page is from the website of an aircraft manufacturer. It gives information about an air conditioning system.

1 Read the text quickly. Is the air conditioning system for a) airport buildings; b) aircraft; or c) hospitals?

2 Read the text again. Find at least five advantages of this system over other systems.

3 Compare your answers with a partner.

Grammar and writing

1 In the text, find and circle at least three comparative adjectives ending in *-er*. What is being compared?

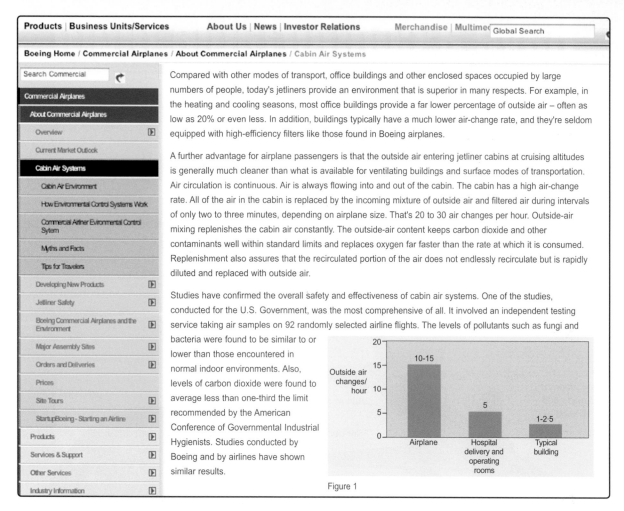

Boeing Home / Commercial Airplanes / About Commercial Airplanes / Cabin Air Systems

Products | Business Units/Services   About Us | News | Investor Relations   Merchandise | Multimed | Global Search

Search Commercial

Commercial Airplanes

About Commercial Airplanes

Overview

Current Market Outlook

Cabin Air Systems

Cabin Air Environment

How Environmental Control Systems Work

Commercial Airliner Environmental Control System

Myths and Facts

Tips for Travelers

Developing New Products

Jetliner Safety

Boeing Commercial Airplanes and the Environment

Major Assembly Sites

Orders and Deliveries

Prices

Site Tours

StartupBoeing - Starting an Airline

Products

Services & Support

Other Services

Industry Information

Compared with other modes of transport, office buildings and other enclosed spaces occupied by large numbers of people, today's jetliners provide an environment that is superior in many respects. For example, in the heating and cooling seasons, most office buildings provide a far lower percentage of outside air – often as low as 20% or even less. In addition, buildings typically have a much lower air-change rate, and they're seldom equipped with high-efficiency filters like those found in Boeing airplanes.

A further advantage for airplane passengers is that the outside air entering jetliner cabins at cruising altitudes is generally much cleaner than what is available for ventilating buildings and surface modes of transportation. Air circulation is continuous. Air is always flowing into and out of the cabin. The cabin has a high air-change rate. All of the air in the cabin is replaced by the incoming mixture of outside air and filtered air during intervals of only two to three minutes, depending on airplane size. That's 20 to 30 air changes per hour. Outside-air mixing replenishes the cabin air constantly. The outside-air content keeps carbon dioxide and other contaminants well within standard limits and replaces oxygen far faster than the rate at which it is consumed. Replenishment also assures that the recirculated portion of the air does not endlessly recirculate but is rapidly diluted and replaced with outside air.

Studies have confirmed the overall safety and effectiveness of cabin air systems. One of the studies, conducted for the U.S. Government, was the most comprehensive of all. It involved an independent testing service taking air samples on 92 randomly selected airline flights. The levels of pollutants such as fungi and bacteria were found to be similar to or lower than those encountered in normal indoor environments. Also, levels of carbon dioxide were found to average less than one-third the limit recommended by the American Conference of Governmental Industrial Hygienists. Studies conducted by Boeing and by airlines have shown similar results.

Figure 1

2 Complete the sentences using your own ideas.

a A good night's sleep is equal to _____ .

b An air ticket to New York costs as much as or more than _____ .

c On average, girls _____ more than boys.

d _____ are as strong as or even stronger than _____ .

e _____ cost even less than _____ .

3 Compare your ideas with another student.

Speaking

Look at Figure 1 above and discuss what the graph shows. Are these statements true (T) or false (F)?

a ____ Some buildings change air less than half as often as hospitals.

b ____ Hospitals change air at least twice as often as aircraft.

c ____ Some aircraft change air up to six times as often as some buildings.

d ____ All aircraft change air more than three times as often as hospitals.

⇨Workbook pages 132–134

Have a comfortable flight

Vocabulary and pronunciation

You are going to listen to a talk about passenger cabin air systems. You will need to be able to identify some compound nouns.

Look at the diagrams and the words in the box. Tick the words you already know. Underline the ones you don't. Use the diagrams to work out what they do.

> heat exchanger   air ducting   cabin floor   filtered air   circular pattern
> floor grilles   overhead outlet   outflow valve   lower lobe

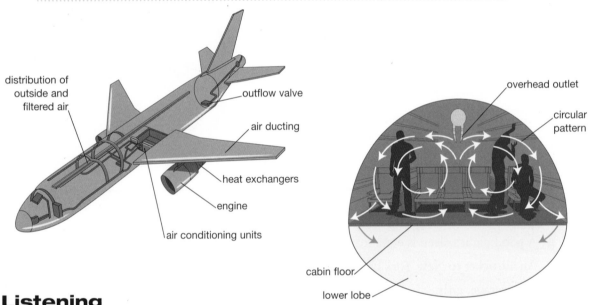

distribution of outside and filtered air

outflow valve

air ducting

heat exchangers

engine

air conditioning units

overhead outlet

circular pattern

cabin floor

lower lobe

Listening

1   Listen to your teacher saying the compounds from Vocabulary and pronunciation. How many syllables are there in each two-word combination? Mark them.

2   Which of the two words in each case carries the main stress?

3   Practise saying the words.

4   🎧 Listen to the talk and raise your hand when you hear the words.

5   Look at the notes below. Discuss with a partner how much you remember between you.

6   🎧 Listen again. This time, complete the notes on the opposite page.

```
NOTES
Engine → cabin
Air comes from _____ first cooled by heat exchangers in struts.
Flows via _____ further cooled _____.
Mixed _____
Comb. outside + filtered air ducted to cabin _____
Inside cabin
Air flows _____ exits _____.
Cabin → exhaust
Exiting air goes below floor into lower lobe.
½ air exhausted _____ valve controls cab. press.
Other ½ mixed w/outside air from _____.
Filters trap _____.
```

7   Compare notes with a partner.

8   Read the tapescript while you listen to check and complete your notes.

Writing

1   What forms of abbreviation are used in the notes in Exercise 6 above?

2   Review your own notes and see if you can make them shorter while ensuring that you can still understand them.

Pronunciation

In each line, one of the underlined vowels has a different sound from the others. Which is it?

1	lower	flow	over	hot
2	valve	drawn	pattern	cabin
3	floor	move	through	cool
4	main	chamber	engine	plane
5	mix	filter	high	grille

Speaking

Work with a partner. Look back at the notes you have for Exercise 2 above. Practise describing the cabin air system, using the notes as guidance.

⇨ Workbook pages 134/135

Vocabulary I

If you cool air, the result is *cooled air*. What is the result if you:

1 pressurise nitrogen? _____

2 compress a fuel and air mixture? _____

3 mix outside and freshened air? _____

4 condition an area? _____

5 heat water? _____

6 expand a gas? _____

Reading and speaking

This is a schematic for a typical Environmental Control System (ECS) unit from a modern small aircraft.

1 **Study the diagram with a partner. Use what you have learnt in Unit 8 to discuss the possible function(s) of each part.**

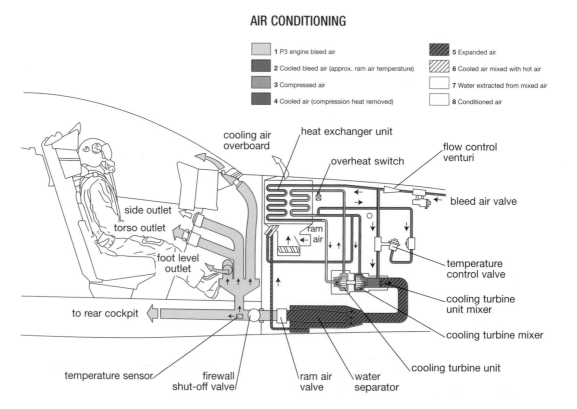

AIR CONDITIONING

1 P3 engine bleed air

2 Cooled bleed air (approx. ram air temperature)

3 Compressed air

4 Cooled air (compression heat removed)

5 Expanded air

6 Cooled air mixed with hot air

7 Water extracted from mixed air

8 Conditioned air

2 Use the labels and key from the diagram to write the correct words next to the definitions.

a Outside air forced into the plane and used only for cooling _ram air_____

b A device used to detect how hot or cold the cockpit air is _____

c Pressurised air which is drawn off from the engine _____

d The open ends of air ducts _____

e A component which dries the air _____

f A two-stage unit which gives off heat to cool air _____

g A small orifice used to direct the flow of air on the canopy _____

h A device used to block the airflow in case of emergency _____

i A two-part rotary component _____

j A shaped tube which controls the rate of airflow _____

Vocabulary II

1 Look at the Language Box. How many other examples can you find in the text and diagram?

2 Discuss with a partner the job of:

a an air distribution nozzle

b an exit grille

c the computer systems maintenance crew

d an ECS

e an emergency package pressurised gas delivery line

> **Language Box**
>
> **Long compound nouns**
>
> Components are described in a particular way, 'in reverse':
> e.g., *firewall shut-off valve*
>
> In final position is the thing itself: *valve.*
>
> Before that, its function, in noun form: *shut-off*
>
> Before that, the sub-system it is associated with: *firewall.*
>
> Similarly, *pressure regulator, flow control venturi.*

Writing and speaking

1 Write a compound noun for:

a a switch which selects the on/off actuation of an air-brake _air-brake activator switch_

b a reservoir which contains hydraulic fluid for the landing gear _____

c a gauge which shows oil temperature _____

d ply which reinforces the tread of a tyre _____

e linkages for the control surfaces at the rear of the aircraft _____

2 Go back to the diagram of the ECS and with your partner describe how the system works and the function of each component.

➡ **Workbook pages 136/137**

Vocabulary I

1 Read the text and label the diagrams below with the underlined vocabulary.

> The thousands of electrical components installed in a modern passenger jet would be useless without something like 250 kilometres of wires, cables and busbars to connect them with each other, as well as with a power source.
>
> The main elements of a cable are the <u>core</u>, which consists of the <u>conductor</u> and its
> 5 <u>insulation sleeve</u>, and the <u>protective covering or sheath</u>. For conductors, the most important factor is current-carrying capacity.

Figure 1:
Cable with round
conductors

Figure 2:
Cable with section-
shaped conductors

Figure 3:
Multi-cored
cable

2 Match the following verbs with the correct noun phrase. There may be more than one correct answer.

verb	noun/noun phrase
dissipate	several layers
carry	heat
make use of	a conductor
insulate	damage
protect something from	a shock absorber
consist of	the conductor from its protective sheath
separate	a current
act as	space

Reading

1 Read the text opposite and circle the expressions underlined in Vocabulary I, Exercise 1 above.

2 Write the title of each section in the text. Choose from this list.

Shape of conductor   Potential problems   Protection   Types of conductor

3 **Now read the text in detail. What types/features of conductors and cables provide the following?**

lightness   efficient heat dissipation   protection from impact   flexibility
good conductivity   space-saving   protection for the cable against the armour
protection against corrosion   a circular cross-section   high current-carrying capacity

4 **Compare answers with a partner.**

Section 1: _____

Copper has a higher conductivity than aluminium, so a copper conductor will carry more current than an aluminium conductor of the same size. However, aluminium is lighter than copper, so it is used in aircraft when possible. The conductor may be solid
5   metal or, if flexibility is required, it may consist of several smaller conductors (strands) fitted inside the insulation sleeve.

Section 2: _____

The greater the cross-section of a conductor, the more current it will carry. A flatter shape has a larger surface area, so it can dissipate heat more easily and carry a larger
10   current. Figures 1, 2 and 3 compare the cross-sections of different types of cable. The sector-shaped conductors make a better use of space than round conductors, and consequently can carry a higher current. A filler is used to fill in spaces to give a circular cross-section.

Section 3: _____

15   In a multi-cored cable, conductors must be insulated from each other. The main materials used for insulation are cotton, PVC, vinyl, Teflon and Rockbestos. The protective covering includes sheathing and armouring. Sheathing protects the cable from hazards, such as water, oil, acid or corrosion. The material used for sheathing depends on the nature of the insulation material. Heavier cable 'armouring' usually
20   consists of steel wire or tape wound around the cable separated from the cores by a layer of bedding material.

Section 4: _____

One problem with all insulating materials is that they can break down at high voltages. In other words, they start to behave like conductors. This breakdown voltage
25   is also affected by the temperature: the higher the temperature, the lower the breakdown voltage. Another is mechanical damage, such as abrasion or impact, especially to heavy-duty cable. Armouring avoids this, and the bedding material prevents the armour from damaging the insulation on the conductor by acting as a shock absorber when the cable is being moved.

⇨ **Workbook pages 138/139**

Vocabulary

Match the terms *voltage*, *current* and *resistance* with their units and definitions.

	unit	definition
voltage	ohms	a steady movement of energy, e.g., the flow of an electrical charge
current	volts	the opposition of something to an electric current passing through it, so that the current changes into heat or another form of energy
resistance	amperes	the difference in electrical potential between two points in an electrical or electronic circuit

Reading and speaking I

1 Read the first part of the text. Which of these does it explain?

a How to use mathematical tables in general.

b How to differentiate between voltage, current, resistance and power.

c How to calculate voltage, current, resistance and power.

> **Ohm's Law**
>
> The type and size of a conductor depends on the voltage, current and resistance of the power in the electrical circuit it is designed for. The relationship between these quantities is expressed by the following two formulae:
>
> 5 $V = I \times R$ and $W = V \times I$
>
> V = voltage (volts), I = current (amps), R = resistance (ohms), W = power (watts)
>
> If you know any two of the four values, you can calculate either of the others by using the relevant formula. To visualise the relationships, look at these six circles. The shaded value is the product or the quotient of the unshaded values, so for example,
>
> 10 V=IR and I=W/V, etc.

> 1   2   3   4   5   6
>
> Example calculations:
>
> Diagram 1 says $V = I \times R$, so if I is 10 and R is 4, then V = 10 x 4, which is 40.
>
> Diagram 3 says R = V/I, so if V is 24 and I is 12, then R = 24/12, which is 2.
>
> Diagram 5 says I =W/V. If W is 18 and V is 6, then I is 18/6, which is 3.

2 Choose a diagram. Tell your partner two of the values from it. He will find the relevant diagram to work out the third. For example:

Student A: V is 45. R is 6. What is I?

Student B: In Diagram 2, I is V over R, so I is 45 over 6. That's 7.5.

Use a calculator if you wish.

Reading and speaking II

1 Look at the chart below and find this information.

a What is the title?

b What are the numbers on the left-hand axis?

c What are the diagonal lines?

d What are Curve 1 and Curve 2?

e What are the four columns on the left?

f What are the numbers along the top and bottom?

2 Now read the second part of the text, which explains the chart. Follow the example calculation on the chart.

A key consideration when selecting a conductor is that the longer the conductor, the greater its resistance. Resistance
5 in the conductor means that there will be a voltage drop at the end. To avoid this, the conductor used must be as thick and/or short as possible.
10 The table above is used to decide which size of wire is suitable for a given length. The larger the wire size number is, the less current it is capable of carrying.

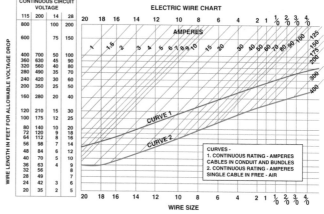

15 For example, assume the requirement is to find the nearest safe wire size for a 30-foot-long bundled cable in a conduit which has to conduct a 28-volt supply to a 1-kilowatt actuator motor. From the formulae above:

I=W/V. Therefore, the current, in amps, is 1000/28 = 35.71.

With this information, go to the Wire Size chart. Look at AMPERES at the top, and
20 go to the value of 35.71 on the diagonal lines. Then follow the diagonal down until it meets Curve 1 (conduit cable readings) at 30 ft according to the fourth column (28 V) on the left-hand axis. The reading is between 8 and 10; the nearest safe wire size is taken as 8 (the lower number is chosen for safety).

➡ **Workbook pages 139/140**

Vocabulary and pronunciation

1 **Can you match the names of the items to the pictures?**

> conduit   grommet   terminal block   cable tie   cable clamp   lug   cable

a   b   c   d   e   f   g

2 **Discuss the questions with a partner. Check that you know the meaning of the words that are underlined.**

a Which item has a <u>loop</u> at the end?

b Which item has a <u>closed ring</u> at the end?

c Which item could be tightened and loosened?

d Which item could cause people to trip and fall if it were not <u>routed</u> correctly or if it had too much <u>slack</u>?

e Which item has <u>holes</u> in it?

f Which item is a channel for holding cables?

g Which item is a rubber <u>seal</u> that protects against dirt and prevents <u>chafing</u>?

3 **Mark the stress in these phrases and practise saying them aloud.**

a you can see from the identification number

b nobody's likely to walk into them

c they haven't left a loop

d I can move it up and down at least a couple of inches

e I'm afraid it'll have to be redone

Reading

New installation or replacement of electrical wires and cables in an aircraft must be done in accordance with certain standards, otherwise the plane cannot be considered airworthy.

1 **Read the maintenance form on the opposite page.**

a Who is the maintenance inspector?

b What is the code number of the form?

c What two areas does this form focus on?

d How do you fill in the different sections?

2 Discuss with a partner what possible problems the inspector could find with each item on the form.

Listening

1 🎧 Listen to Mr Townsend talking about the electrical inspection. Which of the things below have a problem?

  a the ID no.   **b** 6" dia. loop   **c** grommet   **d** drainage holes

2 🎧 Listen again and complete the form.

3 Compare answers with a partner.

Ace Aircraft Ltd
Electrical Maintenance Check

Aircraft and I.D.	Date	Inspector (print name)	Usual signature
ACE 306 – P72	23/5/--	G. TOWNSEND	G A Twnsd
Conductor check	Pass ✓ Fail ✗	Nature of fault	Recommendations
Identification No.			
Conductor size			
Conductor type			
Loop			
Route			
Hole size			
Grommets			
Clamping			
Deflection (slack)			
Cable ties			
Terminals and lugs			
Conduit check	Pass ✓ Fail ✗	Nature of fault	
Position			
Material			
Size			
Bending radii			
Drainage holes			
Deformation			
Form E1			

Speaking

Talk to your partner again. What problems or emergencies might be caused by the things which did not pass the electrical inspection? Use some of the expressions below.

  Loop of wire: *It's needed to allow for …*

  Lugs: *They should be … not … Otherwise, they can …*

  Rubber grommet: *It's needed to stop …*

  Clamp: *It's too … It could …*

  Conduit: …

➡️**Workbook pages 141/142**

Long life

Speaking and vocabulary

1 **Discuss the following questions.**

    a Can you think of five everyday and five industrial uses for batteries?

    b Where have you seen the batteries below?

    c Which batteries have a <u>long life</u>? Which batteries have a <u>low self-discharge rate</u> when they are not being used?

    d What factors can affect a battery's <u>performance</u>?

    e How can they be <u>recharged</u>? Does recharging always lower a battery's <u>capacity</u>?

    f What type of battery is found in cars?

2 **Look at the questions again. Check you understand the differences between the words below.**

    a self-discharge/recharge   **b** performance/life/capacity

Reading I

1 **Look at the text below quickly and answer questions *a–c*.**

    a What aspect of batteries does the text deal with?

    b Which four types of battery are mentioned?

    c Which is the main ambient factor affecting battery performance?

> **Battery life and operation: self-discharge**
>
> Most aircraft are fitted with a heavy-duty onboard battery for the engine starter motor. It is therefore vital to ensure that batteries are kept fully charged and in good condition at all times.
>
> 5   Although self-discharge is normal for all batteries, it will occur at a different rate for different types of battery, as follows:
>
> - Lead-acid: approx. 5% per month or 50% per year.
> - Lithium-ion: approx. 5% in the first 24 hours after charge, then 1% to 2% per month.

10 ● Nickel-based batteries have a relatively high self-discharge rate.

  ● A new nickel-cadmium battery loses approx. 10% of its capacity in the first 24 hours after charge, settling to about 10% per month afterwards.

  ● Nickel-metal-hydride: about 30% higher per month than for nickel-cadmium.

  Note: A critical factor affecting both the useful life and the operation of batteries is
15 ambient operating temperature, since chemical reactions will take place more quickly in high temperatures. Therefore, batteries will perform much better in a warm climate; however, even when in storage and not connected to a load, they will also self-discharge more rapidly.

2 Use the information in the text to mark the following true (T) or false (F). You may need a calculator.

a ____ Nickel metal-hydride batteries self-discharge nearly three times more quickly than Ni-cad batteries.

b ____ Lead acid and lithium-ion batteries have similar self-discharge rates.

c ____ Nickel-metal-hydride batteries have a much higher self-discharge rate than lead-acid batteries.

d ____ Ni-cad batteries are at 20% of their capacity after the first month.

e ____ After two years, a lead-acid battery will have lost about ¾ of its original charge.

3 Check your answers with a partner.

Reading II and speaking

These two graphs show the performance and the self-discharge of typical lithium-ion batteries.

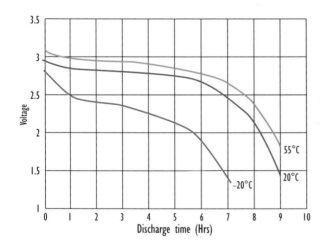
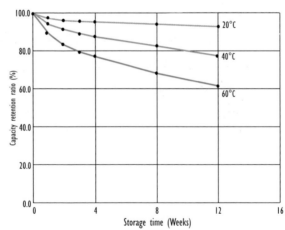

Work with a partner to discuss these two questions.

a How long does it take for the charge to fall to 1.5 V at 20 degrees C?

b What is the capacity after four weeks at 60 degrees C?

➡ **Workbook pages 142–144**

Vocabulary

Choose the correct word from the box to fill the gaps in the sentences below.

> to   out   into   off   up   down   round   over

1 Don't leave your headlights on. You'll run _____ the car's battery.
2 Did you leave the battery to charge _____ overnight.
3 Can you hear the engine turning _____ ?
4 The engine has cut _____ completely.
5 We could try hand-propping: turning the propeller _____ by hand.
6 It's better if you have an alternator fitted _____ the plane.
7 The plane is starting to taxi _____ .
8 Be careful you don't run _____ anything!

Speaking

Work with a partner. Look at the picture below. How do you think this happened?

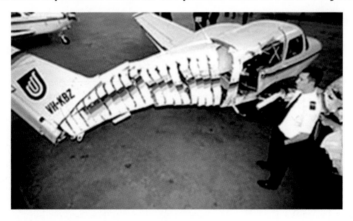

Listening and writing

1 🎧 Listen to the radio news item and find out if you were right.

2 Compare what you understood with your partner.

3 🎧 Listen again and complete the report form on the opposite page.

Form HIR1/a	
HANGAR INCIDENT REPORT: Aircraft damage	
Type and ID of aircraft	
Name of owner	
Other persons involved	
Type of accident (collision, fire, etc.)	
Cause (human/mechanical failure, etc.)	
Description of incident	
Damage to aircraft	
Damage to hangar/equipment	
Personal injury (brief description only; fill in and attach form HIR3 as necessary)	
Signature	Date
To be signed and handed to the Airfield Office within 24 hours of incident	

4 🎧 **Listen to the recording. Are these sentences the same as what you hear? Mark them S (same) or D (different).**

a ____ The engine wouldn't start up on the electric starter motor.

b ____ It's been really hot the last couple of weeks and that makes them run down faster.

c ____ The electric starter motor becomes an alternator and charges the battery up.

d ____ Get an alternator fitted to your plane if you haven't got one.

e ____ I went up four weeks ago so the battery was charged – it was flat.

f ____ We've got two single-engined things. Light aircraft.

5 **Use the tapescript to practise reading the correct sentences aloud, concentrating on stress.**

Language

1 **Look at the Language Box.**

2 **Write the separate events involved in the following sentences. For example:**

I should have taken the battery out yesterday and charged it up overnight. *He didn't charge the battery up.* Anyway, I hadn't, so it didn't work when I tried it. *He tried it and it didn't work.*

a I'd run a couple of electrical checks and discovered it was the battery.

b Yesterday was my first time up since I'd been on holiday.

c The battery had self-discharged – it was flat this morning.

d The plane taxied in because the hangar doors had been left open.

e I hand-propped it easily as I'd done it several times before.

➡ **Workbook pages 145/146**

> ## Language Box
>
> *had* + past participle
>
> The verb form *had* + past participle is often used to refer to an action before the past situation under discussion.
>
> Example:
> The battery *had* self-discharged – it was flat.
>
> He got a bit scared and forgot what I'*d* told him.

Start-up

Vocabulary

Complete each sentence with a word from the box below.

> build up   burn out   snag   become pitted   short circuit   worn out

1 Leads and cables can become tangled and they can _____ .
2 Bearings have a limited life and soon become _____ .
3 When carbon deposits _____, they decrease the efficiency of an engine or motor.
4 When two conductors accidentally touch, they can cause a _____ .
5 Metal surfaces become _____ due to arcing (sparking).
6 Motors and generators can easily _____ if they are not used correctly.

Reading I and speaking

This multi-graph shows the performance of a 24 V starter motor in good condition.

1 **Study the graph and find:**

   a the values scales for each of the lines represented.

   b the units each aspect of the performance is measured in.

   c the amps scale.

2 **Answer these questions.**

   a At 200 amps, what is the efficiency?

   b When the motor is running at 1,400 RPM, what is the current?

   c What is the maximum torque reading?

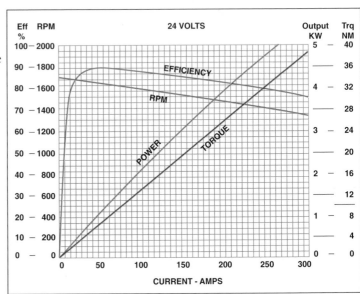

3 **Work with a partner. Ask and answer similar questions to those in Exercise 2 above.**

Reading II

To maintain optimal performance figures, the motor must be kept in good condition: regular inspection and repair are required.

Look quickly at the following text about maintenance. Which of these are dealt with?

> magnets   bearings   shaft length   radio static
> armature   brush wear   start-up and cooldown period

Starter generators are liable to early burn out if they are not used correctly. At start-up, aircraft starter generators draw an extremely heavy current, well in excess of
5 500 amps, which causes them to get very hot during the start sequence. Consult the aircraft operating manual for maximum starting lengths and minimum cooldown periods and ensure that these are adhered to.

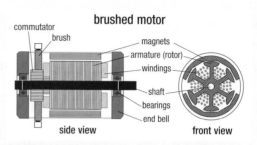

brushed motor

commutator
brush
magnets
armature (rotor)
windings
shaft
bearings
end bell
side view   front view

If the brush leads are not correctly positioned following routine brush inspection or
10 replacement, they can snag on the brush holders and prevent the brushes from contacting the commutator correctly as normal wear occurs. If there are air gaps between the brushes and commutator, electrical arcing (sparking) occurs, resulting in accelerated brush wear and commutator pitting and burning as well as interference with radio equipment. Inspect the commutator for burn marks or pitting and blow out
15 any excessive build-up of carbon with compressed air. Position the leads correctly so that they will not be disturbed when reinstalling the fan cover.

Vibration caused by rough or worn out bearings or out of balance armatures can have a serious effect on nearby components as well as reducing the efficiency and working life of the SG unit. During maintenance, never drop the unit on the drive
20 shaft or put excessive pressure on the bearings or armature. Bearings have a limited life, and a good rule of thumb is to replace them within three years of overhaul.

Radio static is often a sign of the brushes sparking caused by excessive wear, or possibly a shorted armature. It is also caused by failure of the capacitors on the terminal block, which should be checked with a capacitance meter.

Language

The causes of problems.

1   **Find and circle these expressions in the Reading text above (they are in order).**

> are liable to   causes them to   they can   occurs   resulting in
> can have a serious effect on   is often a sign of   is also caused by

2   **Use the expressions above to talk about the results of the following situations/problems.**

   a The brush leads are incorrectly positioned.

   b There are air gaps between the brushes and the commutator.

   c The armature is out of balance.

➡Workbook pages 146–148

Troubleshooting

Look inside any piece of electronic equipment and you will find a board with a number of different components on one side and on the other, printed copper tracks, which are the electrical conductors for these components. This is the *printed circuit board* – PCB.

Speaking and listening

1 **Look at the different representations of PCBs below. With a partner, answer the questions. Which one(s) show(s):**

   a a three-dimensional image?   ____

   b a flow chart?   ____

   c both sides of the board?   ____

   d the holes for leads to be inserted?   ____

   e the layout of the copper-conducting tracks? ____

2 🎧 **Listen to descriptions of the different PCB images and match each one to one of the diagrams and pictures.**

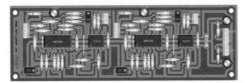

a Component and track layout diagram

b Pictorial view of electronic board

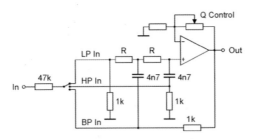

c Electronic circuit schematic

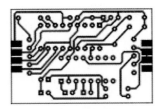

d Printed circuit board track layout diagram

e Block diagram of a radio receiver system (generic)

3 🎧 **Listen again and make notes on a) what it shows and b) possible uses for each representation.**

Reading

Read the following text about methods of fault-finding electronic equipment. Choose one of the words from the box to fill each gap.

> emitted   volume   several   locate   signal   sequence
> inject   instructions   antenna   loudspeaker

Fault-finding

There are several methods of finding faults in electronic circuits.

One method is to use a **flow chart**. This functions by testing
5 various different parts of the system in a logical
_____ to establish where the fault lies, following a series of _____
10 with yes/no outcomes. Here is a simple example for fault-finding a lamp that is not working.

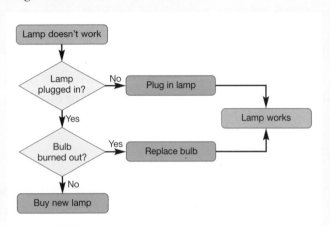

Another method is called the **half split** method.

15 Any piece of equipment, such as a transistor radio, has _____ sequential stages. The signal from the _____ passes through these stages and is finally _____ from the loudspeaker as an audio _____.
Imagine you have one that does not produce an audio output signal. Where does the fault lie? It could be in any one of the stages.

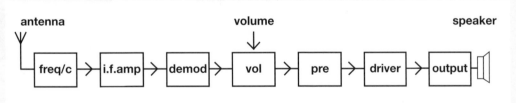

20 The _____ control is about halfway along this chain. If we _____ an audio signal at this point and hear noise from the _____, then we know that all stages and components after this point are OK and the fault lies before this point.

From this one measurement, we have proved that half of the components are OK and that the fault lies in a certain area. Further half split measurements will enable us to
25 _____ the precise stage in which the fault lies.

If we had started at the antenna end and the fault was in the loudspeaker, then we would have wasted a lot more time and effort before we found it.

➡️**Workbook pages 148/149**

Online purchase

Vocabulary and speaking

1   Work with a partner. Try to find these items in the pictures and say what they are used for.

> multimeter   pliers   tape measure   soldering equipment   screwdriver

2   Write the names of any other tools and equipment that your recognise.

3   Mark the stress on the names of all of the items and practise saying them aloud.

Listening

1   🎧 Listen to someone buying some electrical tools and instruments to use on a small aircraft electrical system. Put a tick beside each item he buys.

2 Listen again and make notes explaining the advantages of each of the following items according to the salesman. The notes for the pocket voltage tester have already been completed.

a pocket voltage tester

    v. cheap   115 V dangerous   easy + quick to use – kept in pocket

b analogue multimeter

c tape measure

d solder gun

e socket set

f multi-grip pliers

g single-side cutters and long-nose pliers

Pronunciation

Consonant sound combinations with *l* and *r*

1 Practice these combinations of sounds.

fl	fly fluid	
pl	explain pliers plus	
bl	cable blowtorch	
cl	click clamp	
tr	electric strip try	
cr	crimping across	
pr	product protect	
dr	hydraulic draw screwdriver	

2 Think of other words, or find words in the Word lists, which contain these sounds.

3 Add some of the words to the column on the right and practise them.

➪ Workbook pages 149–151

Lesson 9
Power systems

Vocabulary

1 Complete each dictionary definition below with a word from the box.

> DC   inverter   generator   AC   online   EPU

a _____ (*n*) equipment for supplying electricity to an aircraft using a separate battery unit stored on the ground: short for **External Power Unit**.

b _____ (*adj*) in operation, working, functioning. **be ~** be operational: (computers) be working using the Internet. (systems in general) **bring ~** put into operation.

c _____ (*n*) rotating machine that converts mechanical energy into electrical energy, e.g., **starter ~**.

d _____ (*n*) electrical device that converts direct current into alternating current.

e _____ (*n*) alternating electric current that changes direction with a regular frequency, as in domestic mains electricity.

f _____ (*n*) a direct electric current that flows steadily in one direction.

2 Discuss the following questions with a partner.

a What systems or functions in an aircraft require an electricity supply?

b How do you think power is supplied to them?

Reading

1 Read the text and try to find answers to the questions in Vocabulary Exercise 2.

2 Find and circle the following numbers in the text.

> 32 V   400 hertz   115 V   700   26 V   40 amp/h   24 V

3 Read the text around the numbers to find out what they refer to.

> **POWER SYSTEMS**
>
> **DC power system**
>
> DC power is three-way: primary DC power is supplied by the dual-role starter-generator (SG); a nickel-cadmium battery forms the secondary DC power source;
> 5   external DC power can be supplied via an EPU connection. When a connected EPU is brought online, the generator and battery are isolated from the system.

Starter-generator

The SG is a dual-role unit mounted on the engine accessory drive housing. SG output is supplied to the aircraft systems at a regulated 28 V DC. During flight, or with the
10 aircraft on the ground with the engine operating, all DC power is obtained from the SG, which also trickle-charges the aircraft battery. If the SG malfunctions, or is inadvertently switched off, the battery is automatically brought online. If the generator output exceeds 32 V, a voltage regulator automatically disconnects the system from the generator and reconnects it to the battery. With the engine shut down, DC power
15 is obtained from the battery or an EPU.

Battery

The 24-volt 40 amp/hour nickel-cadmium battery is located in the rear fuselage. It only comes online when there is no output from the SG or the EPU. In order to reduce the risk caused by battery overheating, there is an over-temperature warning
20 system which lights up the BAT HOT caption on the cockpit annunciator LCD display panel when the battery temperature reaches 700.

External power supply

A minimum of 24 V and maximum of 28 V power can be supplied to the aircraft via the EPU connection. When external power is supplied to the aircraft DC system, the
25 battery is isolated from the system to prevent the possibility of it discharging through the EPU.

Most of the onboard equipment is powered by the DC system.

AC power system

Some of the flight instruments and avionics require an AC supply. The AC generation
30 system consists of two transistorised static inverters, which each convert the 28 V DC supply to 26 V AC and 115 V AC at a frequency of 400 hertz. Only one inverter is online at a time, the other serving as an emergency backup.

4 Read the text again and make notes, using the headings below in the blue boxes.

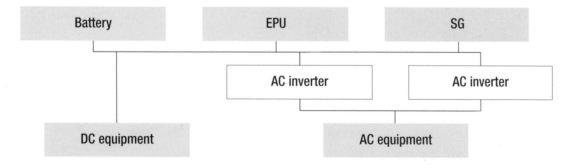

5 Compare your notes with a partner's.

Speaking

Using your notes only, give a brief talk to your colleagues on the power system of the aircraft.

⇨ **Workbook pages 151–153**

Vocabulary

Match the words below with a set of words in the table that can go with it.

> a circuit   operated   electrical   a fuse   a switch
> equipment   a fault   current   socket

a	heavy (*adj*) AC (*adj*) regulated (*adj*)	*current*
b	power (*adj*) multiple (*adj*) electric (*adj*)	
c	distribution (*n*) device (*n*) component (*n*)	
d	detect (*v*) repair (*v*) diagnose (*v*)	
e	multiple (*adj*) remote (*adj*) electromagnetic (*adj*)	
f	DC (*adj*) onboard (*adj*) high-voltage (*adj*)	
g	melt (*v*) replace (*v*) remove (*v*)	
h	complete (*v*) break (*v*) overload (v)	
i	automatically (*adv*) manually (*adv*) battery (*n*)	

Reading and speaking

1   **Read this description of busbars and complete the text with the expressions supplied.**

Busbars are an essential part of the
_____ system, acting as a
common _____ for groups of
different components. They are similar to
5   multiple sockets. They carry a
_____ and are designed to
dissipate heat quickly. They are made as
_____ or hollow tubes of
copper or aluminium. These shapes have the highest ratio of _____ so
10   that the _____ of heat escapes.

> electrical distribution   heavy current   wide flat strips
> surface area to volume   maximum amount   connection point

2 **Work with a partner. One of you uses Texts A and B; the other Texts C and D.**

a Read your texts.

b Take turns to explain the components without using the text itself; take notes as your partner speaks.

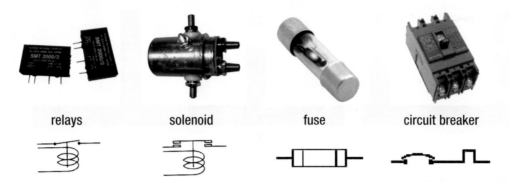

relays   solenoid   fuse   circuit breaker

A Solenoids are electromagnets which move a metal rod in and out of a central coil. They operate equally well with both AC and DC current. They are used to actuate small mechanical devices, valves and switches.

B Circuit breakers are devices which will break a circuit very quickly in an
5   emergency situation. They can then be reset when normal operation is resumed. They can operate automatically, manually or both. They can be actuated by mechanical pressure, heat or electrical current.

C Relays are DC-operated electromagnetic switches which use low voltage and low current to enable heavy-current and high-voltage equipment to be switched on
10   and off from a distance. They also enable multiple switching to be performed from just one remote switch.

D Fuses are conductors which are designed to melt if a specified current passes through them. The melting, which is caused by the current flow, breaks the connection and prevents the circuit being overloaded. They must then be removed
15   and replaced once the original electrical fault has been detected.

Writing

1 Individually, write your notes from Exercise 2 above into a brief text.

2 Compare your texts with the originals. The information content is important; the exact structure of the original text is not.

3 Exchange texts and look for possible errors of sentence structure.

⇨ **Workbook pages 153–154**

light plane

747

fighter

Speaking

What information do you think would be useful to a pilot when he is taking off, landing or cruising? Make a list and then compare it with a partner.

Reading

1 Look quickly at the text below and compare it with your ideas.

2 Read the text and the list more carefully, and mark each item in the list as follows:
    E = external conditions, **PE** = performance, **W** = warning, **PO** = position.

A pilot needs to be aware of a lot of information on take-off and landing, and in flight. He must understand fully at all times:

- the position of the aircraft;
- the aircraft's performance;
5 - external conditions which could affect the safety and efficiency of the aircraft;
- potential emergency situations.

In addition to the onboard instruments, there will be communications equipment which enables the pilot to speak to other aircraft and to air traffic controllers, and most planes nowadays also carry a flight data recorder, which keeps a record of what 10 happens to the plane during a flight. The following is a list of key pilot information on different aspects of the flight:

• Altitude	____	• Cabin pressure drop	____
• Height	____	• Attitude	____
• Air speed	____	• Trim	____
15 • Landing gear status	____	• Engine speed	____
• Collision	____	• Distance from departure point	____

- Wind speed _____
- Ground speed _____
- Cabin pressure _____
20 - Vertical speed _____
- Distance to destination _____
- Fire _____
- Air temperature _____
- Local traffic _____
25 - Hot battery _____

- Fuel supply _____
- Oil pressure _____
- Generator voltage low _____
- Battery status _____
- Electrical system failure _____
- Cabin temperature _____
- Weather conditions _____
- Fuel low _____

Vocabulary

1   Match the halves of the word combinations without looking back at the text. There may be more than one possibility for each.

landing	battery
wind	drop
ground	departure point
cabin	supply
pressure	gear
distance from	failure
fuel	traffic
weather	speed
electrical system	speed
local	pressure
hot	pressure
oil	conditions

2   Check that you know the meaning of each combination.

Pronunciation

1   Look at these words. Underline the long vowel sounds.

take   sound   fly   blow   noise

2   How many words in the list in Vocabulary Exercise 1 can you find which contain these long sounds?

3   Add more words from the Reading text.

4   Practise the correct pronunciation of the expressions in the list in Vocabulary Exercise 1 above, with correct stress and long sounds.

➡ Workbook page 155

Speaking

1   Try to identify each of the cockpit instruments in the pictures. Choose from this list.

gyro compass   altimeter   attitude indicator

turn coordinator   vertical speed indicator   air speed indicator

a   b   c   d   e   f

2   What information does each one give the pilot?

Listening and writing

You are going to listen to a flying instructor giving a lesson about the six basic flying instruments.

1   🎧 Listen to the recording. Fill in the blank instrument panel with the correct instrument layout by writing A–F in the circles.

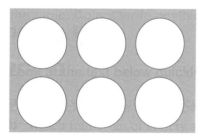

2   Draw a table in your notebook and make notes on each of the six instruments. Remember to use note form only, including block capitals. Use the following layout.

	what it tells pilot	why needed	how it works
air speed indicator			

3   Compare and discuss your notes with your partner's.

4   🎧 Listen again to check. You can follow the tapescript if you wish.

Pronunciation

1 Mark the stressed words in these sentences.

    a The layout of the panel is likely to be the same in any small plane you fly.

    b You can't look at an instrument and then switch off.

    c How does it work?

    d You'll shake it and break it.

    e This measures the aircraft's rate of climb or descent.

2 Listen to the recording and check your answers.

3 Practise saying all the sentences aloud.

Language

If

> **Language Box**
>
> *if*
>
> The word *if* is used with words like *will/may/might* in sentences which express possible results.
>
> Example:
> *If the ambient temperature is high, the battery **might** self-discharge more quickly.*
>
> However, *if* is also used where there is no case of possibility.
>
> Example:
> *The lights come on if you press this button.*
> *If you need help, just ask.*

1 **Look at the Language Box. Which of the following are a statement of a <u>truth</u> and which express the result of a <u>possible</u> situation? Mark them T or P.**

    ____ All gases expand if they get hot.

    ____ There'll be some turbulence if we meet bad weather.

    ____ It makes it easier to adapt if you change to flying a different plane.

    ____ If there was no turn coordinator, the pilot might slip sideways during turns.

2 **How many sentences using *if* can you find in the tapescript? Are they T or P?**

3 **Now complete these sentences in your own words.**
For example:

If you don't go fast enough, ... *the aircraft might stall and start to fall.*

    a If you don't calibrate your altimeter before take-off,_____.

    b If the battery becomes overheated,_____.

    c If a magnetic compass is moved around a lot,_____.

    d If you can't see clearly out of the cockpit,_____.

    e If you sit in the cockpit of any light aircraft,_____.

4 **Compare your sentences with your partner's.**

➪**Workbook pages 156/157**

Speaking and vocabulary

1   **Work with a partner. Match the pictures with the labels in the box.**

> digital watch   electricity meter   magnetic compass   digital multi-meter
> spring balance   digital thermometer   mercury thermometer

 a
 b
 c
 d
 e
 f
 g

2   **Discuss these questions.**

   a What do they do?

   b Which are analogue, and which digital?

   c What are the differences between the analogue and digital thermometers below?

   d What exactly do the terms *analogue* and *digital* mean?

Reading and writing

1   **Think about digital and analogue instruments. Write questions by expanding the following.**

   a What/used/for?_____

   b What/mean?_____

   c What/advantages?_____

   d What/disadvantages?_____

2   **Student A: read Text A; Student B: read Text B. Find the answers to the questions you have just written and make notes. Remember to use note form: include abbreviations, miss out articles, etc.**

> Text A
>
> Originally, the instruments and electronic equipment installed in aircraft were analogue devices. An analogue is a kind of copy of something. For example, if electric current increases by 10%, the pointer of a moving coil ammeter will move 10%
> 5   further along the dial: the movement of the pointer is an analogue of the increase in the current. In other words, there is a direct relationship between the input to the meter and the output on the display, which can be very useful. Another important feature of analogue devices is that they work in a continuous way, e.g., an analogue

voltmeter has a needle which moves smoothly from one position to another. Analogue
10 displays also show the complete range of readings of the quantity being measured so
that the operator can see where a reading is on the scale between a maximum and
minimum position.

However, analogue devices are more likely to be affected by extremes of temperature
and pressure, as well as requiring lubrication. They are also more prone to
15 malfunction and more difficult to read accurately than digital displays. Just as some
analogue equipment has a digital display (a home electricity meter, for example), it is
possible for digital devices to have analogue displays.

Text B

Nowadays, a lot of control, measuring and communications equipment is digital. This
means that it works in a different way from analogue devices and usually has a display
which shows only digits or whole numbers. The greatest advantage that many digital
5 instruments have is that they do not have any moving parts: they are solid-state devices.
This means they are usually less fragile and they do not wear out or require lubrication.
They also tend to be less affected by extremes of temperature and pressure, and
mechanical shocks. For this reason, they are often more suitable for use in aircraft.

Another advantage of digital displays (e.g., watches) is that they are easy to read
10 quickly. Digital displays are also more accurate, and they are often cheaper and
simpler to make and repair than analogue displays, although they need a power source
of some kind. However, they do not always show the operator the maximum and
minimum on the scale being measured, or show gradual changes in the same obvious
way which analogue displays do. It is possible for digital devices to have analogue
15 displays, but technical instruments which have a digital display have the very practical
advantage that it is normally easier to get a correct reading from them.

Speaking

1 **Work in pairs, A and B. Ask and answer questions about each other's texts, using your notes.**

2 **Discuss which you would prefer to read, analogue or digital displays, and why, given the reasons in the texts.**

Language

Tendency

1 **In Texts A and B, find these expressions.**

> not always   be prone to   tend to   be affected by   usually

2 **Which of the expressions is used in a negative sense, and which positive?**

➭**Workbook pages 157/158**

Know-how

Speaking

Avionics (aviation electronics) systems are now an important and integral part of the design, repair and maintenance of aircraft navigation, communications, radar, instruments and computers that control flight manoeuvres, engine performance and environmental systems.

What specific areas do you think a modern avionics technician needs to know about? For example, specialist tools, using the manual correctly.

Reading I

1   **Look quickly. Are your ideas from Speaking above mentioned in *a–e*?**

   a Practical applications of engineering science and technology, including materials, principles, techniques, procedures and equipment for the design and production of avionics equipment and systems.

   b Schematic, layout, electromechanical drawings of avionics components and systems.

   c The application of arithmetic, algebra, geometry, calculus and statistics to electronic and electrical equipment.

   d Circuit boards and electronic equipment.

   e Hardware and software, including applications and programming, processors and chips.

2   **Match each of the five sentences with one of these three headings.**

   i Computers and Electronics   **ii** Mathematics   **iii** Engineering and Technology

Vocabulary

1   **Look at the verb and noun combinations in column 1. Choose another noun from column 2 that can also go with each verb.**

   a interpret drawings   a schedule   _____

   b coordinate work   faults   _____

c	install wiring	equipment	_____
d	set up support systems	malfunctions	_____
e	fine-tune an engine	assemblies	_____
f	carry out an inspection	data	_____
g	adjust equipment	problems	_____
h	diagnose faults	tests	_____

2 Add one more noun in column 3 which could go with each verb.

Reading II

1 Find and underline the verb-noun combinations from Vocabulary Exercise 1 in the following list of activities of the avionics technician.

The avionics technician needs to be able to:

- set up and operate ground support and test equipment to carry out functional flight tests of electrical and electronic systems.
- calibrate, regulate and fine-tune avionics equipment for optimum performance.
5 - interpret flight test data in order to diagnose malfunctions and systemic performance problems.
- test and troubleshoot instruments, components and assemblies, using circuit testers, oscilloscopes and voltmeters.
- interpret and refer to technical drawings.
10 - adjust, repair or replace malfunctioning components or assemblies.
- keep clear records of maintenance and repair work.
- coordinate work with that of engineers, technicians and other aircraft maintenance personnel.
- install electrical and electronic components, assemblies and systems in aircraft.
15 - connect components to assemblies such as radio systems, instruments, magnetos, inverters and in-flight refuelling systems.
- assemble components such as circuit boards, switches, electrical controls and junction boxes.
- fabricate parts and test aids as required.

2 Talk about activities which the avionics technician has to do. Which activities do you think are the most a) challenging and b) interesting? Use the verbs and nouns from the Vocabulary exercise.

⇨Workbook pages 159–161

Speaking

The way that a diagram is drawn can make a difference to how well we understand how something works.

Work with a partner. Look at these two circuits.

    a What do they show?

    b What is the difference between them?

    c Which one do you find easier to understand?

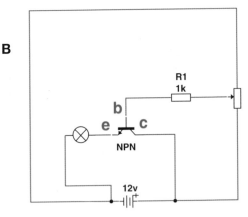

Reading

Read the two texts below.

    a Match them to diagrams C and D.

    b What kind of diagram is each one?

> 1 This drawing shows the relationship between the different types of component found in a gyro compass. It shows that the gyro system consists of electrical, mechanical and electronic components. All three types are shown together, so it is referred to as an electromechanical diagram. This can make the drawing more complex, but an overview like this means that it is easier for a trained avionics technician to understand how the whole assembly works.

> 2 As well as the functional relationship between components in electronic circuits, their relative position and proximity can affect the performance of the equipment. This is particularly important for high-frequency and radio circuits of the kind that are installed in modern aircraft. To minimise interference, layout diagrams can be drawn to show where components should be positioned on the chassis or on circuit boards, as in this DC motor.

C

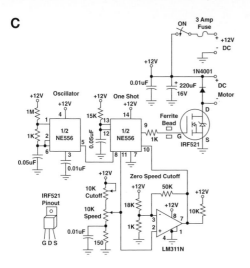

D

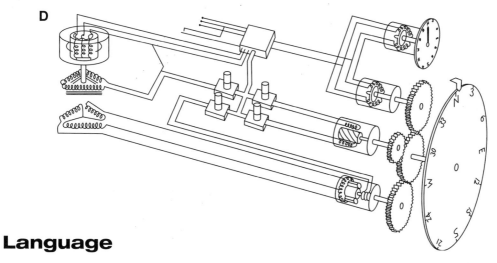

Language

Make

Look at the Language Box. Then make sentences beginning with *a–i*, using *make it*.
For example:

> Poor lighting **makes it** hard (for the operator) to see the job properly.

- **a** Poor lighting
- **b** Standard instrument layout
- **c** Multi-grip pliers
- **d** Digital displays
- **e** Bad weather
- **f** A cockpit turn coordinator
- **g** Workplace safety regulations
- **h** Internationally recognised symbols
- **i** Exploded drawings

➭ **Workbook pages 161–163**

> ## Language Box
>
> *make it* + adjective
>
> Look at the example:
> *The drawing **makes it easier to** understand how the whole assembly works.*
>
> In this case, *it* does not refer to the drawing, but to the understanding. This pattern is often used with the following adjectives: *make it easy/easier to* + verb, *make it difficult/more difficult/harder to* + verb, *make it likely/more likely/less likely that* + clause, *make it possible to* + verb.

Speaking

Look at these two printed circuit board (PCB) layout drawings. What is the difference between them?

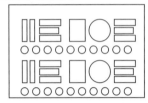 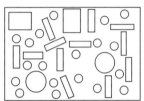

Vocabulary

You are going to listen to an instructor talking to two trainees about troubleshooting and repairing circuit boards in the workshop. First answer these questions.

1 Who or what ...

    a develops a fault? _____   **b** repairs a fault? _____   **c** traces a fault? _____

2 Which of these troubleshooting terms are good news for the technician and which bad?

> fault comes and goes   cracked   repairable   faulty   badly arranged   OK
> badly made   broken away   fitted neatly   intermittent fault   loose
> mishandled   tidy   put under stress   not soldered properly   insecure connection

Listening

1 ⌂ Listen to the conversation in the recording.

    a How many of the things in Vocabulary Exercise 2 above are mentioned? Tick them as you hear them.

    b Make a note of each problem and possible solution in the table below. Compare your ideas with a partner.

problems	solutions
1 2	

2 🎧 **Listen again. Complete Rasheed's notes from the training session with Mr Patel.**

PCBs

Layout

v. imp to keep board _____ and _____ .

Direction of comps. should be _____ because

_____ .

2 basic kinds of fault:

_____ and _____ .

Specific faults, e.g., _____ , insecure connections

Causes, e.g., _____ , _____ or _____ .

Faults sometimes intermittent, i.e., _____ .

Repair

1st check visual – use _____ and magnifying glass – check

for visible _____ + bad connections.

2nd check PCB: connect to a _____ and press on

_____ .

3rd check indiv _____ .

If comp. faulty, always _____ . (?)

Pronunciation

1 🎧 **Look at the Skills Box. Listen to these sentences in the recording and mark the links between words.**

a If the board is put under a lot of stress, it might crack.

b One of the problems with insecure connections is that the fault is often intermittent.

c Throw it in the rubbish.

d Press down with a pencil in different places.

e If the board itself seems OK, put very gentle pressure on each component.

f Here is the circuit diagram and a list of the readings.

2 **Practise saying the sentences aloud.**

➡ Workbook pages 163/164

Skills Box

Listening for key items

You will need to practise identifying key items in a stream of speech. These could be numbers, names or ideas.

Linked speech

In spoken English, words are often linked smoothly together. This means that there is no break before a word beginning with a vowel, for example:

to ⌒ each ⌒ other

the board ⌒ in front ⌒ of you

there ⌒ aren't ⌒ any ⌒ obvious cracks

As always, you need to be ready to hear this kind of linked speech and to use it to make your English easy to understand.

Vocabulary

Look at these words and phrases.

> a solvent cleaner   a short circuit   foreign matter   a solder sucker   grease
> a heat sink   reinstallation   protective lacquer   a small gap

1 Mark the ones you know with a tick and the ones you think you might know with a question mark.

2 Work with a partner to check the ones you don't know.

Reading

1 Read the instructions in Exercise 3 below and mark any of the words/phrases from the Vocabulary exercise that you find.

2 What is this text?

   a General instructions for repairing PCBs

   b General description of PCBs

   c General electronics maintenance

3 Supply a suitable verb to complete each instruction. Choose from the following.

> remove   solder or unsolder   paint or spray   clip on
> avoid   ensur   remove   brush or blow   trim

   a _____ bending, flexing or twisting the PCB while you are working on it by holding it in a suitable stand.

   b _____ off any loose particles of foreign matter before checking for PCB faults.

   c _____ any grease from the area to be worked on with a suitable aerosol solvent cleaner.

   d _____ connections as quickly as possible so that the board and components are exposed to the minimum of heat.

   e _____ a heat sink whenever possible to dissipate excess heat.

   f _____ any excess solder with a solder sucker so that conductors are not short-circuited and holes are not filled.

g _____ that all components have a small gap to allow ventilation airflow between the component and PCB.

h _____ the connecting wires of horizontal components so that they are of equal length and as short as possible.

i _____ conductor repairs with protective lacquer and allow them to dry thoroughly before reinstallation of the PCB.

4 Work with a partner to match each instruction *a–i* to one of these pictures.

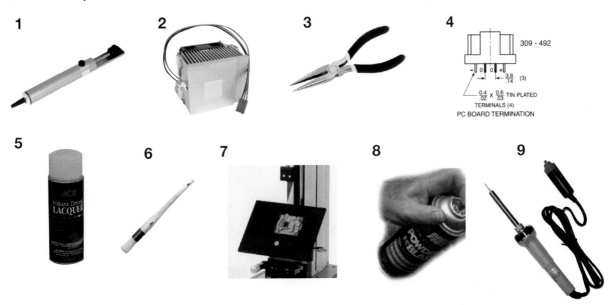

1   **2**   **3**   **4**

309 - 492

3.6 (3)
.14

0.4 x 0.6 TIN PLATED
.02 .03 TERMINALS (4)
PC BOARD TERMINATION

5   **6**   **7**   **8**   **9**

Language

Degree

1 Why are these instructions important? Look at them again and discuss with a partner the reason for each requirement. Use *might*, *can*, *could* or *may* to add a reason to each instruction. For example:

> If component wires are too close together or too far apart, they might be put under stress. In addition, the wires might not fit properly in the holes – they could be pushed in too far, or not far enough.

2 Discuss these questions with a partner.

In the maintenance workshop in general, what should operators do:

a as much as possible?

b whenever possible?

c thoroughly?

d as little as possible?

e at all times?

f under no circumstances?

g only under supervision?

➡ **Workbook pages 164–166**

Vocabulary and speaking

The picture below shows most of the different radio aerials or antennas fitted to a Boeing 737.

1 Here is a list of communications devices. Work with a partner.

> traffic alert and collision avoidance system   air traffic control
> global positioning system   automatic direction finder   high frequency
> distance measuring equipment   very high frequency   emergency locator transmitter

a Match the names of the devices with the initials on the picture.

Antenna locations

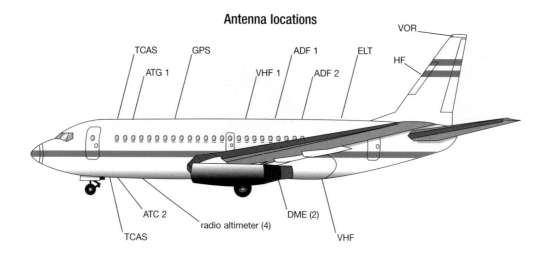

b Discuss what you think each device might be for.

2 Look only at the picture. How many can you remember?

Listening

You are going to listen to part of a lecture on radio navigation aids.

1 Work in a group of three. Each member of the group should look at one group of words below and check that you understand them. Then explain them to your partners.

- transmitter, receiver, quadrant, dashes and dots
- tower, controller, path, navigation
- radio signal sent, echo received, frequencies, chart

2 Say all the words aloud after the teacher.

3 🎧 Listen to the lecture. Put the following, all described by the speaker, into the order in which they were invented.

airport control towers   two-way air-to-ground radio

radar   Very High Frequency Omni-directional Radio

four-course radio range

4 🎧 Listen again. Label the correct diagrams for the 1929 Four-course Radio, ATC and RADAR on page 250.

Language and writing

Linking ideas

1 Complete these paragraphs using *because, still, although, and, but* or *so*.

- A pilot knew whether he was flying toward or away from a Morse transmitter
  a _____ the signal would get louder or more faint as he flew. He could not
  know his exact heading, **b** _____ he knew which corner of the square he was
  nearest to, **c** _____ he had an idea of his direction **d** _____ the
  distance to the transmitter.

- Even after control towers became common, pilots **e** _____ had to use visuals to
  inform the tower of their position. Radio navigation aids were not accurate enough for
  controllers to find aircraft on a chart, **f** _____ pilots had to report known
  landmarks to the tower.

- **g** _____ it is important to keep a safe distance between aircraft in flight,
  RADAR was an important invention. **h** _____ it does not allow a controller to
  guide a pilot on landing, it is useful in planning safe flight paths **i** _____
  advising pilots of the relative position of other aircraft.

2 Work with a partner to link these ideas.

a Workshops can be dangerous. Safety procedures are essential.

b Digital displays are very useful. Analogue displays have advantages.

c Radial engines are easier to cool than inline engines. They are also smoother.

d Humidity is important for ATC. It affects weather conditions. Pilots need to know about it.

➡️ Workbook pages 166–168

Vocabulary

1 **How many nouns, adjectives and adverbs can you make from these verbs?**

> acquire   transmit   receive   track   inform
> navigate   activate   locate   measure   equip

2 **Work with a partner and write sentences using some of the words you have found.**

Language

Question forms

Expand the following into questions.

1 what / kind / clocks / used ?

2 how / unit / activate ?

3 what / GPS unit / measure ?

4 how / ELT units / identify ?

5 how / information / reach / RCC ?

6 what / unit / look like ?

7 what / 'acquisition' ?

8 what / MCC and RCC / stand for ?

9 how / satellites / identified ?

10 system / accurate ?

Reading

1 **Work in groups, A and B. Group A read the GPS text; group B read the ELT text.**

a Read your own text only and choose which five of the questions in Language above relate to it.

b Answer the questions in note form.

c When answering questions, give as much detail as a you can using your notes. Now take it in turns to ask your partner the other five questions.

GPS

The Global Positioning System (GPS) is becoming the primary means of navigation worldwide. The system is based on satellites in a continuous grid surrounding

5 the Earth at an altitude of approximately 20,200 km, each equipped with an atomic clock set to Coordinated Universal Time (UTC) at Greenwich, England, called 'Zulu time' or 'military time', which means it is written in four digits, e.g., 16.30 rather than civilian 4.30 p.m.

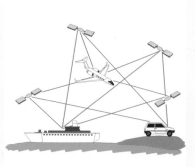

10 The GPS units in the aircraft, or even in a pilot's hand, find the nearest three satellite signals in a process called 'acquisition'. Once it has identified the satellites' locations from their acquisition codes, it measures the time delay between transmission and reception of each satellite's radio signal and calculates the distance to each satellite, since the signal travels at a known speed. The pilot then knows his latitude and
15 longitude, accurate to within one metre.

The accuracy of GPS is affected by atmospheric conditions, including humidity and reflections of signals from buildings and geographical features. The navigation message from a satellite is also transmitted only every 12.5 minutes, so the data is slightly out of date, which may cause up to 2 m of inaccuracy. In coming years, GPS
20 will be made even more precise and include ground transmitters at the ends of runways to aid landing.

ELT

Most modern commercial aircraft are required to carry an Emergency Locator Transmitter (ELT). ELTs are tracking transmitters: when activated,
5 they send out a signal that allows the unit to be located to within 100 metres, anywhere in the world, by a low-Earth-orbit satellite overhead. This information is transmitted by the satellite to a receiver system on the ground at a Mission
10 Control Centre (MCC), and the location is passed on to a Rescue Coordination Centre (RCC), which is responsible for the search and resue operation.

Most ELTs are brightly-coloured, waterproof, fit in a cube about 30 cm square and weigh 2–5 kg. The units have a useful life of 10 years, operate across a range of
15 conditions (–40°C to 40°C), and transmit for 24 to 48 hours. They are required to transmit on 406 MHz from 2009; modern 406 MHz ELTs transmit in bursts and remain silent for a few seconds in order to conserve transmitter power.

The ELT becomes active when, in an emergency situation, the pilot selects an emergency radio frequency manually, or self-activates automatically when the aircraft exceeds a
20 certain force in landing, called the g-force, during a crash, helping to save the lives of injured pilots and crew who may be unable to activate it manually.

Speaking

1 **From what you now know, describe the information in your partner's text to him. He/she will correct you if necessary.**

2 **Read your partner's text to confirm your ideas.**

➡ **Workbook pages 168/169**

See-through metal

Vocabulary

1 Work with a partner. What is the difference between the following?

AC and DC current

a crack and a void

a magnetic field and an eddy current

permeability and durability

corrosion and dirt

a conductor and a coil

something that is sound and something that is faulty

2 Practise the pronunciation of these words with your teacher and partners.

Listening

You are going to listen to a conversation about inspecting the internal condition of metal parts.

1 First, look at these pictures and discuss with a partner what they show.

2 🎧 Listen to the conversation and number the pictures in the order they are mentioned.

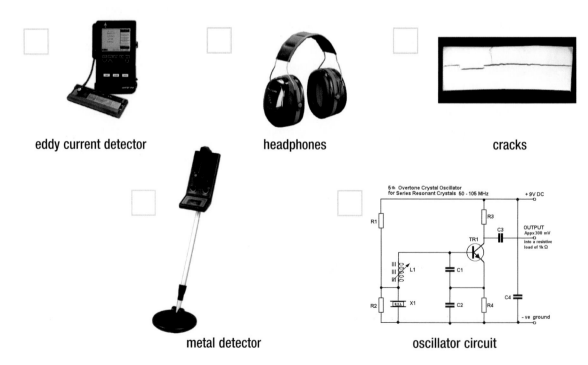

eddy current detector

headphones

cracks

metal detector

oscillator circuit

3 Read the notes below. Then listen again and correct them as required.

Metal detecting instruments

- Use oscillator circuit to create DC voltage in coil.
- Coil connected to headphones. Coil has elec. mag. field.
- Elec. mag. field in detector induces current in any object on ground and creates elec. mag. field in obj.
- Instrument detects other object's shape.
- Not exactly used for airport security.

Eddy current inspection instruments

- *Used to check for faulty, cracked instruments in plane, e.g., landing gear/engine mountings.*
- *Large cracks always under paint or on dirty surfaces.*
- *Eddy current flows through damage easily.*
- *First take reading (refar...?) from sound of metal.*
- *Compare test reading with first reading. If same, no prob. if different, fix part.*

Pronounciation

1 Look at the Skills Box. Then group these words under the correct consonant sound.

signal	something	through
that	direction	there
engine	gear	piece
frequency	operate	mobile
excessive	change	dangerous
check	bit	current
various	physics	

Skills Box

Pronunciation

English has pairs of consonant sounds, for example, *s* and *z*, where the mouth is the same for both, but one sound uses only air and the other uses the voice.

Example:
f and *v* *k* and *g* *p* and *b*
ch and *dge*

It is important for understanding that you pronounce and hear the difference clearly.

get /g/	cool /k/	affect /f/	very /v/	bolt /b/	open /p/	watch /tʃ/	large /dʒ/	three /θ/	this /ð/

2 Think of at least two other words for each column.

3 Practise saying the words aloud.

⇨Workbook pages 169–171

Vocabulary

Complete the definitions with an item from the box.

| relationship | mandatory | appropriate | certify | overhaul | Authority | factor | regulate |

a _____ (n) element affecting a situation or a result. *Overall weight is an important ~ in engine design.*

b _____ (adj) in accordance with a rule. *All new operators must complete the ~ training course successfully.*

c _____ (n) official organisation responsible for a certain area of industry. *These are the new rules from the Workplace Health & Safety ~ .*

d _____ (v) to control in accordance with rules. *We need to ~ the use of dangerous equipment.*

e _____ (v) to mark or register something to show it has been checked and fulfils requirements for use. *An inspector must ~ rebuilt engines before they can be reinstalled.*

f _____ (adj) suitable and correct for a particular case. *Anyone reporting an accident has to fill in the ~ forms.*

g _____ (n) the way in which people or systems are connected and work together. *The ~ between speed, distance and time can be expressed as S=D/T.*

h _____ (v, n) **1** (v) to disassemble, check and repair completely. **2** (n) the process of disassembling, checking and repairing something. *The report says that the system needs a complete ~ .*

Reading

1 **Look at the diagram opposite. Work with a partner. Discuss what each of the three icons might represent.**

2 **Read the text quickly and see if you were right. Label the three icons in the diagram.**

3 **Read the text again and answer the questions below.**

a What can be done to improve the safety of each element in aircraft?

b What is the job of the ICAO, FAA and EASA?

c Who is responsible for filling in forms?

4 **Complete the text with the words from the Vocabulary section. You will need to change the form of some of the words.**

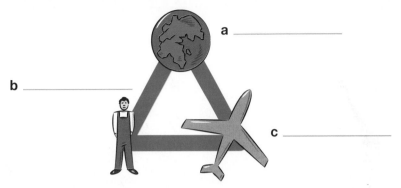

a _____

b _____

c _____

The safety of an aircraft depends on a triangle of _____. Each one is important, and the _____ between them is also important. The external environment cannot be controlled, although extreme environmental conditions can be avoided. However, control of people and machines can be _____. People
5 who work with aircraft, such as pilots and maintenance personnel, can only do so if they have the _____ certificate from the correct _____.
Similarly, machines – airframes, systems and all component parts – have to be checked and _____ in order for the plane to be considered airworthy. Some of the main authorities that control airworthiness are:

10 ICAO = International Civil Aviation Authority

FAA = Federal Aviation Administration

EASA = European Aviation Safety Agency

It is the job of the maintenance personnel to ensure that the _____ regular inspections, _____ , repairs and replacements are carried out
15 correctly and that all the necessary forms are correctly filled in.

Speaking

Look at the forms on pages 251 and 252.

1 **Which relate to:**

a airworthiness?

b faults?

c overhaul?

d non-certified components?

2 **Work in pairs or small groups. Choose one form each and find out:**

a when it would be used.

b what kind of information would be required in each section.

3 **Show your forms to your partner(s) and explain them.**

⇨ Workbook pages 172/173

Lesson 2
We all make mistakes

Vocabulary

1 **Decide whether the pairs of words have a similar or different meaning in each case.**

a do up/fasten

b insufficient/inappropriate

c precise/accurate

d deactivate/switch off

e spare/useless

f reassemble/rebuild

g live/current-carrying

h tags/labels

i bolt/ screw

j unsuitable/wrong

2 **With a partner, discuss any differences in meaning.**

Language and speaking

There are two main kinds of mistake that are made by maintenance personnel:

A They do things they shouldn't do.

B They fail to do something.

1 **Here are 15 maintenance errors which could cause equipment failure. Decide whether they are in Category A or Category B. Then compare answers with a partner.**

a _____ wrong kind of hydraulic fluid put into brake system

b _____ insufficient lubrication of motor shaft

c ___ equipment cover panel not refastened

d ___ 5-amp capacity cable put into 10-amp circuit

e ___ screw hammered into position

f ___ tools left in work area

g ___ connectors not securely done up

h ___ oil-based grease used on rubber components

i ___ maintenance record documents not completed

j ___ spanner used instead of torque wrench

k ___ spare solder left on PCB repair

l ___ reassembly instructions ignored

m ___ live cable cut with insulated side-cutters

n ___ no warning tags on unfinished repair

o ___ bolt refitted without locking washer

2 **Complete these sentences with the correct form of the verbs given. There may be more than one possibility.**

a That's very hot – if you _____ it, you _____ yourself. (touch, burn)

b What _____ if you _____ the volume of a gas? (happen, increase)

c The project is well advanced now. If we _____ now, we _____ a lot of time and money. (stop, lose)

d This is an electric fire. What _____ if I _____ a water extinguisher on it? (happen, use)

e You should avoid dropping an electric motor – the drive shaft _____ misaligned. (become)

f If I _____ this copper wire into the acid, you _____ a reaction begin to take place. Watch carefully please. (put, see)

3 **Compare sentences with a partner. Make sure that the sentences make sense and, between you, correct any sentences that do not.**

4 **Look again at the list of maintenance errors in Exercise 1. With a partner, explain how the mistakes described could affect an aircraft system. Use *should ... because ... if ...***

For example:

Screws shouldn't be hammered in. If you hit a screw head with a hammer, you would probably damage it, and you would definitely damage the material you were screwing into. That would make future maintenance difficult and you might cause a dangerous defect.

➡ **Workbook pages 173–175**

Language Box

Hypothetical conditions

The word *if* is used with words like *would/could/might* in sentences which express hypothetical situations, e.g., *If you left your tools in the work area, someone else would probably take them.*

Getting it the right way round

Vocabulary

The majority of maintenance mistakes are related to the reassembly and reinstallation of assemblies, subassemblies and single components. This is because it is usually much easier to disassemble something than to put it back together.

1 **Some words from the text above have been added to this table. Work with a partner to fill in the rest.**

verb	undo it	do it again	process
			maintenance
	disassemble		
		reinstall	
		put back together	

2 **Now try these.**

verb	undo it	do it again	process
mount			
			fitting

3 **Here are some expressions for positioning components and subassemblies.**

  a Match them to their opposite meanings in the box.

> the wrong way round   the other way round
> upside down   on the wrong side   leave out

   i the right way up

  ii put in

 iii on the right side

  iv this way round

   v the right way round

  b Which expression *i–v* means:

_____ correct orientation?   _____ vertical orientation?

_____ incorrect orientation?   _____ horizontal orientation?

Reading and speaking

Work with a partner.

1 Look at this simple food mixer. Can you see what mistake could be made when reassembling it?

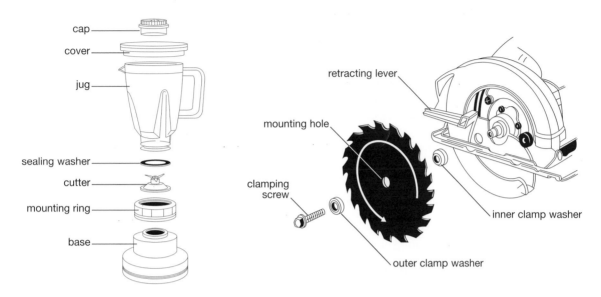

2 Now look at the power saw. There are several different mistakes you could make putting it together. Can you find them?

3 Finally, look at the diagram below of a much more complex assembly.

a What is this assembly?

b Discuss the correct procedure for reassembling it.

c Talk about mistakes that could be made during reassembly. Use the following expressions.

You might do X when you should do Y.

You could forget to do Z.

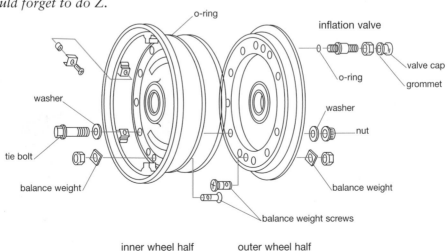

➡️**Workbook pages 176/177**

Listening and speaking

You are going to listen to a maintenance manager speaking.

1   **Work with a partner. Think of five features of documents that make them easier to read and understand, e.g.,** *headings on forms*, *diagrams in manuals*.

2   🎧 **Listen once to the manager. What is he talking about?**

   a a particular maintenance procedure

   b maintenance documentation style

   c reading skills for engineers

3   🎧 **Listen again. Make a note of the main problems and solutions the speaker describes.**

Reading and writing

1   **Look at the instructions opposite and answer these questions.**

   a What job are these instructions for?

   b What equipment do they refer to? Choose from the pictures below.

2 **What is wrong with the way the notes are written, according to the maintenance manager?**

1. THE APPLIANCE SHOULD INITIALLY BE ISOLATED FROM THE MAIN SUPPLY OF ELECTRICITY IN ORDER TO OBVIATE THE RISK OF DANGEROUS ELECTRIC SHOCK HAPPENING.

2. DISMOUNT UNIT FROM WALL SO AS TO FACILITATE SERVICING PROCEDURE OF UNIT.

3. YOU SHOULD CAREFULLY REMOVE THE FOUR SCREWS FASTENING THE COVER OF THE UNIT AND LIFT THE COVER TO REVEAL THE APPLIANCE. YOU MUST REMOVE THE FAN.

4. EMPLOYING A BRUSH, BLOWER OR ANY OTHER TYPE OF SUITABLE CLEANING DEVICE OR MATERIAL, ANY ACCUMULATIONS OF DIRT, DUST, FLUFF OR FOREIGN OBJECTS SHOULD BE REMOVED, WHILST BEING VERY CAREFUL AT THE SAME TIME TO ENSURE THAT ANY DAMAGE TO INTERNAL CONNECTIONS, SAFETY CUT-OUTS, THERMOSTATS OR OTHER INTERNAL COMPONENTS WITHIN THE HEATING APPLIANCE IS AVOIDED.

5. APPLY A FEW DROPS OF A36 LIGHT MACHINE OIL TO THE BEARINGS AT EACH END OF THE MOTOR, MAKING SURE THAT YOU DON'T CONTAMINATE OTHER AREAS OR COMPONENTS WITH LUBRICANT.

6. THE COVER OF THE APPLIANCE MUST NOW BE REPLACED BACK ON THE BASE AND THEN RESECURED WITH THE FOUR FIXING SCREWS. YOU MAY NOW REMOUNT THE APPLIANCE IN ITS POSITION ON THE WALL.

3 **Rewrite the instructions in a more appropriate form for a maintenance manual. Work in three groups, A, B and C.**

a Group A: rewrite them using imperatives.

b Group B: rewrite them using the active voice.

c Group C: rewrite them using the passive voice.

4 **Exghange your work with the other groups and read their versions. Discuss the questions below.**

a Are the instructions consistent?

b Which version do you prefer?

⇨**Workbook pages 177–179**

Speaking

One of the important functions of maintenance personnel is to carry out inspection and testing without damaging or destroying the equipment, known as NDI and NDT.

With a partner, discuss the following questions.

   a What do you think NDI and NDT stand for?

   b Think of at least five things which personnel involved in NDI and NDT should or shouldn't do.

   c What are the possible advantages and disadvantages of this kind of maintenance?

Vocabulary

You are going to read about problems with mechanical equipment on a vehicle.

1 **Put the words and phrases in the box in the correct category in the table.**

> wear   deep scratches   sanding   wiping   brake pad   standing water
> wire brushing   rough ground   disc faces   brake lining   rust   industrial pollution

damage to assembly	environmental factors	assembly parts	maintenance actions

2 **With a partner, practise pronouncing the words correctly.**

3 **With a partner, discuss the following questions.**

   a What does this diagram show?

   b How does it work?

   c Which part of the assembly might be affected by each of the factors in Exercise 1?

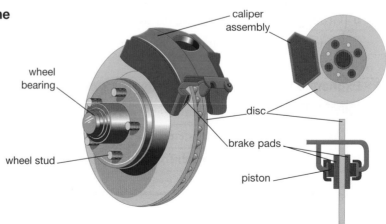

Reading

You are going to read an example manual from an American aircraft wheel and brake manufacturer.

1 Look quickly through the following text and underline the words and phrases from Vocabulary Exercise 1.

Under average field conditions, a brake disc should give years of trouble-free service. However, unimproved airfields, standing water, heavy industrial pollution or infrequent use of the aircraft may necessitate more frequent inspection of discs to prolong the life of the brake lining
5 on the brake pad. Generally, the disc faces should be checked for wear, grooves, deep scratches, excessive general pitting or coning of the brake disc. Coning beyond 0.015 inch (0.381 mm) in either direction would be cause for replacement. Single or isolated grooves up to 0.030 inch (0.76 mm) deep should not be cause for replacement, although general grooving of the disc faces will reduce lining life.

10 Discs are plated for special applications only; therefore, rust in varying degrees can occur. If a powder rust appears, one or two braking applications during taxi should wipe the disc clear. Rust allowed to progress beyond this point may require removal of the disc from wheel assembly to properly clean both faces. Wire brushing followed by sanding with 220 grit sandpaper can restore the braking surface for continued use.

15 pitting = small holes in the surface of the metal
coning = change in the shape of the disc caused by excessive heat

2 Now look at this table. It summarises the information in the text, but it is not completely accurate. Look again at the text and correct the table as necessary.

brake discs – maintenance procedure	
maintenance schedule	
frequent use	more frequent inspection
dry conditions	normal inspection schedule
airborne contamination	less frequent inspection
tolerances	
coning	max. 0.381" (0.015 mm) in either direction
grooves	min. depth 0.030" (0.76 m)
wear	see manual Fig. "A" dim. A2 for max. thickness
actions	
light rusting	clean disc twice before taxiing to wipe
extensive rusting	remove wheel. Apply wire brush, then 022 sandpaper

➡ **Workbook pages 179–181**

Reading and speaking

1 **What meaning do all of these verbs share?**

> ensure   verify   confirm   ascertain   make sure

2 **Look at the following text. With a partner, discuss these questions.**

a What are the people in the pictures doing?

b Why is it important that they do this?

c Can you suggest some reasons why they might miss things during this job?

d What items/systems might a pre-flight visual check involve?

3 **Read the following text and tick any of your ideas that are mentioned.**

Before a flight, it is mandatory for the captain or the copilot to make an inspection of the aircraft, whether any maintenance has been done or not. This pre-flight inspection includes:

1. visually inspecting for obvious structural damage;

2. ensuring that all access doors are secure;

3. verifying that landing gear locking pins have been removed;

4. checking that the control surface locks have been removed;

5. ascertaining that there are no fuel or oil leaks;

6. making sure that engine air ducts are clear of foreign objects;

7. verifying that tyres are correctly inflated and in acceptable condition;

8. confirming that fuel, oil and other liquid systems have been properly serviced;

9. going through pre-flight checks of instruments and indicators in the cockpit.

Listening

Unfortunately, flight crews don't always spot problems. You are going to listen to some reports from aircrew on the results of poor maintenance work.

1 Revise the three forms of the following verbs. Underline the eight verbs that do not end in ~ed in the past and past participle forms. For example, <u>tell</u> (*told*).

take off	land	call	discover	fly	accept
try	check	say	be	do	notice
move	fail	seem	push	arrive	suspect
have to	miss	hand over	give	tell	indicate

2 Look at the Skills Box. Check that your pronunciation of the verbs is correct.

3 🎧 Listen to four reports from aircrew and decide which items on the list in the text from the Reading and speaking section are being referred to.

Writing and language

Study the Language Box.

1 Look at the sentences below. Decide if the result of the situation is in the past or the present.

a If there had been a warning flag on the pin, *someone would have noticed it*.

b *It would be easier to believe if* the captain himself hadn't missed it.

2 Write sentences expressing these past conditions and their results.

a I didn't ask for an authorised fuel delivery document, so I didn't know how much fuel there was in the tanks.

If I had asked ———————————, I would ———————————.

b I'm annoyed because maintenance didn't do a good job.

I wouldn't ——————————— if maintenance had ———————————.

c The heavy traffic meant that we landed at another airport.

We wouldn't ——————————— if the traffic hadn't ———————————.

d I didn't spot the damage with my flashlight, so now I'm in trouble.

If I ———————————, I wouldn't ———————————.

e The locking pin was in position, so we weren't able to bring up the landing gear.

⟹**Workbook pages 181–183**

In safe hands

Vocabulary

1 **Complete these verbs.**

a op _ r _ t _   **d** l _ c _ t _   **g** c _ nc _ l

b pr _ v _ d _   **e** ext _ nd   **h** r _ t _ t _

c pr _ v _ nt   **f** pr _ tr _ de

2 **Think of at least one noun form for each verb.**

3 **Put the words in the box into the correct column of the table.**

caster	hydraulic	chassis	clearance	mechanism	control	access
secure	operational	excessive	jack	handle	accidental	sway

noun	verb	adjective

Reading and writing

Maintenance cannot be done properly without suitable access equipment. Sometimes it is necessary to move personnel, tools and equipment to high-up, inaccessible parts of an aircraft. It is sometimes necessary to lift the aircraft itself.

You are going to read about two types of maintenance equipment.

1 **Work in pairs.**

Student A: Read Text A about the Rotozoom access platform.

Student B: Read Text B about the AJJ series lift system.

2 **From your text:**

a find out, and make a note of:

- equipment type
- operation
- dimensions
- load capacity
- means of transportation
- safety features

b note any of the words from the Vocabulary section above which are in it, and the larger expressions or phrases in which they are used.

Speaking

You and your partner are going to talk about your texts.

1 Write a question for each of the categories in Reading and writing Exercise 2a.

Example:

What type of equipment is it?

2 With your partner, ask and answer each other's questions and make notes of your partner's answers.

3 Student A: Using only your notes from Exercise 2b, present your piece of equipment to your partner.

Student B: Correct any mistakes and tell your partner if he/she leaves out any information.

Then swap roles.

4 Which piece of equipment would be more suitable for:

a use under a low aircraft?

b inspecting airliner windows?

c a load of 800 kg?

d transportation by a single person?

A

Rotozoom access platform

This is a four-wheeled chassis which carries a main servicing equipment section, a rotating and elevating mechanism and a work cage. A remote control unit, normally located on the work cage, is also provided. The remote control unit can be operated from an external position by means of an extension lead. Operational safety is provided by switches which operate in close proximity to adjacent structures and cancel further movement of the platform.

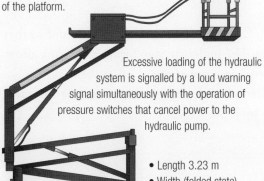

Excessive loading of the hydraulic system is signalled by a loud warning signal simultaneously with the operation of pressure switches that cancel power to the hydraulic pump.

- Length 3.23 m
- Width (folded state) 1.93 m
- Height 1.81
- Weight 3,915 kg (8,613 lb)
- Maximum lifting capacity 230 kg (506 lb)

B

AJJ Jet & Turbo-Prop Aircraft Lift System ... Lift with Confidence

Today's jet and turbo-prop aircraft are too valuable to consider using anything less than the AJJ Series Lift System from Meyer Hydraulics. For over 35 years, Meyer Hydraulics has manufactured aircraft jacks that continue to set the standard for safe, dependable lifting. This custom line of professional lifting equipment handles aircraft weighing up to 32,000 pounds with the security and features one should demand in an aircraft jack.

Features...

(A) ANTI-SWAY ANTI-SNAG GUIDANCE SLEEVES

(B) CONVENIENT HOLE SPACING

(C) POSITIVE SAFETY LOCK Spring loaded safety pin "pops" into the slide tube to prevent accidental lowering.

(D) LARGE-DIAMETER ROUND FRAME TUBE

(E) SPRING-LOADED CASTERS Six 3" casters provide easy transporting.

(F) LOW-HEIGHT LEG DESIGN Provides maximum gear door clearance with one section completely free of any protrusions. The three 26" legs create a wide stance, insuring stability.

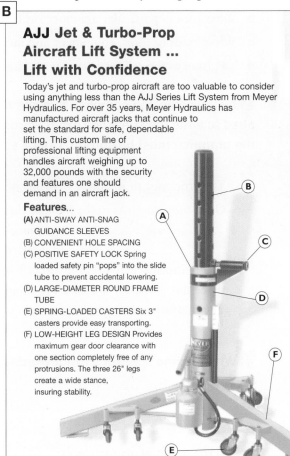

➡ Workbook pages 184/185

Taken out of service

Listening

You are going to listen to a talk on hangar equipment.

1 **Look at this picture of a hangar.**

    a What kind of work would you expect to be carried out there?

    b What jobs in particular are more easily done inside than outside?

    c What kinds of tools, equipment and facilities would be necessary? List as many as you can.

    d 🎧 Listen to the talk and see if your ideas are mentioned.

2 🎧 **Listen again and complete this summary of information from the talk. You do not need to write the speaker's exact words, but the information should be the same and the grammar must be correct. Check your answers with a partner.**

> The efficiency of a hangar depends on _____.
> Once the aircraft is in the hangar, the _____ wraps around so that
> _____ immediately. The undercarriage lifting platforms allow
> _____ undercarriage equipment without _____.
> 5 The platforms also level the aircraft with the access docking before maintenance starts.
> Once this is done, the aircraft is supported by jacks _____.
> The platforms can then be lowered for _____.

3 🎧 **What are the speaker's exact words in these cases? Listen again and write what he says.**

    a _____ should have the facilities you can see here.

    b Heating, ventilation _____.

    c Overhaul, heavy maintenance _____.

    d Works to all parts _____ uninterrupted.

4 🎧 Follow the tapescript as you listen to the recording to check your answers.

5 🎧 Practise saying the sentences, copying the speaker as closely as you can.

Vocabulary

1 Look at the tapescript and underline these expressions referring to the running of a maintenance hangar.

> depends on   provided within the hangar   so as to enable   immediately
> The provision of X allows Y   It is also used to   The platforms can then be lowered
> can continue to be carried out uninterrupted

2 Complete the following table by choosing the correct option in each case.

adjective noun	depends on	to + verb noun
noun to + verb	allows	noun to + verb to + verb
adjective noun	is used to	verb verb + ing
verb noun	requires	noun adjective
adjective noun	provides	noun to + verb
noun to + verb	enables	noun to + verb adjective
to + verb noun	can be	past participle noun

Speaking

Look at the list of hangar facilities mentioned in the speaker's talk.

1 With a partner, discuss why each one is a necessary feature of a well-designed hangar.

- heating, ventilation and air conditioning
- lighting, including emergency lighting
- main, sub-main and small power supply
- fire detection and alarm systems
- fire protection systems
- domestic and process water services
- process ventilation
- compressed air
- lightning protection and main earthing
- energy management

2 Choose one of the areas and explain its importance to a new partner.

➡ Workbook pages 186/187

Equipment MRO

Vocabulary and speaking

Faulty equipment is frequently responsible for damage to aircraft and injuries to personnel.

Work with a partner to complete Exercises *a–c*.

   a Discuss how such damage and injuries might occur.

   b Label the diagram of a trolley jack, using the words and phrases in the box.

> release valve (handle knob)   saddle
> front wheel   handle   foot pedal   rear castor

   c Answer these questions.

      i How do you think you use the jack?

      ii Which parts might need lubrication?

      iii How might you change the hydraulic fluid?

      iv How would you store the jack?

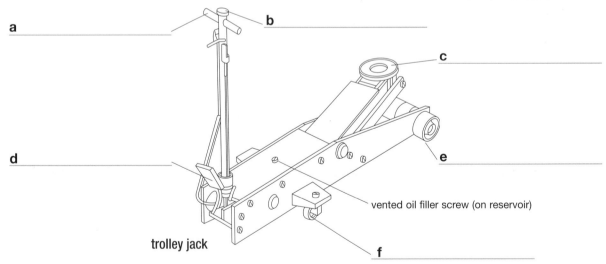

a _____   b _____

c _____

d _____   e _____

vented oil filler screw (on reservoir)

trolley jack

f _____

Reading

Look at the maintenance instructions for the trolley jack below.

   1 **Add one of these headings to the corresponding section opposite.**

> Lubrication Replacing oil Storage Choosing the correct oil Adding oil Inspecting and cleaning

1 _____

Important: Use only a high-grade hydraulic jack oil. Avoid mixing different types of fluid and never use brake fluid, turbine oil, transmission fluid, motor oil or glycerine. Incorrect fluid can cause failure of the jack and the potential for sudden and immediate loss of load.

2 _____

   a)   Set jack in its upright, level position with saddle fully lowered. Locate and remove vented oil filler screw.

   b)   Fill with oil until ³⁄₁₆" above the inner cylinder as seen from the vented oil filler screw hole. Reinstall the vented oil filler screw.

3 _____

For best performance and longest life, replace the complete fluid supply at least once per year.

   a)   With saddle fully lowered, remove the vented oil filler screw.

   b)   Lay the jack on its side and drain the fluid into a suitable container. (Note: Dispose of hydraulic fluid in accordance with local regulations.)

   c)   Fill with oil until ³⁄₁₆" above the inner cylinder as seen from the vented oil filler plug screw. Reinstall the vented oil filler screw.

4 _____

A periodic coating of light lubricating oil to pivot points, axles and hinges will help to prevent rust and ensure that wheels, castors and pump assemblies move freely.

5 _____

Periodically check the pump piston and ram for signs of rust or corrosion. Clean as needed and wipe with an oily cloth. (Note: Never use sandpaper or abrasive material on these surfaces.)

6 _____

When not in use, store the jack with saddle fully lowered.

---

2 **Mark these procedures correct (C) or incorrect (I) according to the instructions in the Reading text.**

a ____ Always use a clean cloth to clean the jack.

b ____ Do not use mixed oils. This can cause complete failure of the equipment.

c ____ Consult the maintenance manual for instructions on disposal of old fluid.

d ____ Lower the saddle when you have finished with it.

e ____ Apply light lubrication to moving parts to prevent corrosion.

f ____ When changing oil, the jack should be fully raised.

g ____ Place a suitable container under the bottom of the jack to drain old oil.

h ____ Maintain a level of approximately ³⁄₁₆" of oil.

⇨**Workbook pages 188/189**

You are going to listen to a conversation about the 'job file' on an aircraft.

Vocabulary and pronunciation

1 **With a partner, decide which of the words in each group is the odd one out. Explain why.**

a personnel   supervisor   authorised person   manager

b record (*v*)   document (*v*)   file (*v*)   write (*v*)

c signed off   okayed   renewed   authorised

d work report   airworthiness certificate   maintenance report   job report

e discrepancy   remedy   problem   fault

2 **These words are from the conversation.**

a Mark the stress on each word.

b 🎧 Listen and check your answers.

i	supervisor	**v**	connectors	**viii**	generator
ii	damaged	**vi**	electrics	**ix**	battery
iii	ordered	**vii**	structural	**x**	replaced
iv	signed off (*v*)				

3 **You are going to hear sentences containing the words from Exercise 2.**

a Look at the Skills Box.

b 🎧 Listen. When you hear a word, write the sentence number beside the word in the list in Exercise 2.

c Look at the tapescript and practise saying the sentences yourself.

Listening

1 **With a partner, look at the job form and discuss:**

a why work is being done on this plane.

b how many technicians have worked on the plane.

c how many categories have been inspected.

d exactly what information might go in the blank spaces in the Discrepancy section of the form.

> **Skills Box**
>
> **Sounds in context**
>
> Words are usually easier to understand when they are heard alone than when they are spoken in a sentence.
>
> It can be difficult to hear weak words, e.g.,
> *has: He has finished the job.*
>
> and to hear linked words, e.g.,
> *the last⌢time I flew*

Fielding Pro-Am Airservice
Maintenance hangar report

OBS: 100-HOUR OVERHAUL

AIRCRAFT	(a)		OWNER D. GREENHILL
Category	Discrepancy		Action
Engine and fuel	Check noisy L/H fuel pump		(b)
Airframe	Inspect and report any defect to owner before doing any repairs		(c)
Hydraulics/controls and landing gear	(d)		Refilled damper with hydraulic fluid & replaced all ring seals
Electrics	(e)		Overhauled SG. connectors renewed
Avionics and instruments	Magnetic compass faulty		(f)

Engine and fuel – Technician/Tradesman		Supervisor/Authorisation	
Signature	Lutfi Tarhoni	Signature	(g)
Name	LUTFI TARHONI	Name	(h)
Date	20/08/2003	Date	(i)
Airframe – Technician/Tradesman		**Supervisor/Authorisation**	
Signature	B Higgins	Signature	John Maddox
Name	(j)	Name	(k)
Date	21/08/2003	Date	22/08/03
Hdrlcs and Ctrls – Technician/Tradesman		**Supervisor/Authorisation**	
Signature	B Higgins	Signature	(l)
Name	(m)	Name	(n)
Date	21/08/2003	Date	(o)
Electrics – Technician/Tradesman		**Supervisor/Authorisation**	
Signature	M L Armstrong	Signature	John Maddox
Name	M L ARMSTRONG	Name	(p)
Date	22/8/03	Date	22/08/03
Avionics & Inst. – Technician/Tradesman		**Supervisor/Authorisation**	
Signature	M L Armstrong	Signature	John Maddox
Name	M L ARMSTRONG	Name	(q)
Date	22/8/03	Date	23/08/03

2 🎧 **Listen to the conversation.**

  a How many people speak? Who are they?

  b When will the plane be ready?

  c Is the owner satisfied with the report?

3 🎧 **Listen again and complete the maintenance hangar form. With a partner, check that your notes contain the right information and are appropriately written. Use the parts of the form already filled in as a guide.**

➡ Workbook pages 190–192

Emergency oxygen supply: Review

Speaking and vocabulary

You are going to read the section of a manual on oxygen supply.

1 **Discuss these questions with a partner.**

    a Why do some aircraft need an oxygen supply for the pilot?

    b When might a pilot require emergency oxygen?

2 **An oxygen cylinder is a cylinder that contains oxygen. A delivery tube is a tube that delivers something. What are the following?**

> a contents gauge   a trip lever   a charging connection
> an operating cable   a manual operating handle   a bolted clamp strap
> a rigid supply tube   a tell-tale wire   an operating mechanism

3 **In each of the multi-word nouns in Exercise 2, which word is stressed? Listen to your teacher's model. Mark the stress and practise saying them.**

Reading

1 **In the text below, circle each item in the list in Exercise 2 above.**

2 **The text from the manual is quite technical. Read it again more slowly and underline the parts that you understand. Leave what you don't understand for the moment.**

3 **Compare with a partner how much you underlined. How much do you understand between you?**

> The emergency oxygen system (EOS) is fitted to the ejection seat and supplies the pilot with oxygen for a limited period during ejection, or on failure of the main oxygen supply during normal flight.
>
> The system comprises an emergency oxygen set, the necessary supply tubing, a trip lever (1)
> 5 mechanism [diagram lower left] and a manual operating handle (2) and linkage.
>
> The emergency oxygen set comprises an oxygen cylinder (3), secured to the seat by a bolted clamp strap (4), and fitted with a regulator assembly. This consists of a capillary tube for controlling the flow, a break-off tube, an operating cable

connected to the trip lever (1), a flexible rubber delivery tube (5), a charging

10 connection (7) and a contents gauge (8). A tell-tale wire (9) is secured around the operating mechanism (6) housing and the trip lever assembly. When broken, it indicates that the set has been operated. The charging connection (7), mounted on the opposite side to the contents gauge (8), incorporates a non-return valve so that a charging hose may be removed from the connection and a blanking union fitted

15 without loss of oxygen. The contents gauge (8) provides a visual check of the cylinder contents. The end fitting of the flexible delivery tube (5) from the regulator is connected by a bayonet connector (10) to a rigid supply tube (10) routed across the upper rear of the seat.

When ejection is initiated, the emergency oxygen supply is automatically turned on by

20 the trip lever (1) as the seat rises. The oxygen supply continues until man/seat separation. In the event of main oxygen failure, emergency oxygen supply is initiated by pulling upward on the emergency oxygen manual operating handle located on the seat left side.

WARNING! IT IS ESSENTIAL THAT ALL EQUIPMENT USED WITH OXYGEN IS

25 KEPT SCRUPULOUSLY CLEAN. THE PRESENCE OF OIL OR GREASE IN CONTACT WITH HIGH-PRESSURE OXYGEN INTRODUCES A SERIOUS RISK OF FIRE AND MAY RESULT IN EXPLOSION. REMOVE ALL TRACES OF OIL, GREASE OR ANY ORGANIC MATTER.

4 Add the vocabulary numbered 1–10 in the text to the diagram.

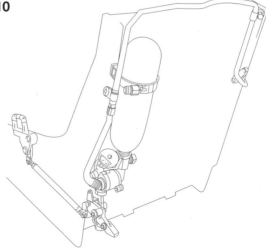

Writing

Look again at the language used in the warning in the text. Choose one of the subjects below and write a similar warning.

- use of PPE
- seat belt inspection
- extinguishing fires
- checking tyres

Use some of the language in the box.

> It is essential that …   It is important to …   Always make sure that …
> ensure …   remove …   check that   may result in / cause …

➡ **Workbook pages 193/194**

Speaking

In Unit 7, you looked at fires and fire extinguishers. Work with a partner.

1 **Try to do the following from memory.**

    a   Match the extinguishers with the classes of fire.

> dry powder
> CO_2
> halon
> water

    b   List the types of materials that are involved in each case.

    c   Sketch the three-part 'triangular' process that keeps a fire burning.

2 **Check your answers in Unit 7.**

3 **Write some advantages and disadvantages of each type of extinguisher in the table below. Think about flexibility of use, the environmental effects of using them, materials they contain, cost, weight and any other factors.**

advantages	disadvantages

Vocabulary

You are going to listen to the results of a fire test. You will need to understand and recognise the following words.

> dust   ignite   glue   smother   backing foam   mess   upholstery fabric

1 **Complete the definitions with a word from the list.**

    a   _____ (*n*) material, usually cloth or leather, used to cover chairs and seats

    b   _____ (*v*) 1. catch fire, begin to burn 2. set fire to

    c   _____ (*v*) join with adhesive; stick (together). ~ (*n*) an adhesive

    d   _____ (*v*) cover completely, e.g., a fire, person, so that air cannot reach it

e _____ (*n*) collection of extremely small dry particles forming a layer on horizontal surfaces

f _____ (*n*) insulation made of plastic filled with many small bubbles of air to make it lighter

g _____ (*n*) disorder, untidiness. a ~: a situation in which there is no order or organisation

2 Practise the pronunciation of the words with your teacher.

Listening

1 🎧 Listen to this description of a test of the three types of extinguisher. Are any of your ideas from Speaking Exercise 3 mentioned?

2 🎧 Listen again and correct the information in the table.

Fire test

fuel: *upholstery, foam, kerosene*

	halon	CO_2	dry chemical
extinguished Y/N	*Y*	*N*	*Y*
no. of squirts required	*22*	*4*	*7*
time	*minimum: ~2–3 secs*	*8–12 secs*	*4 secs*
reignition Y/N	*Y*	*N*	*N*
approx. amount used (%)	*40%*	*100%*	*33%*
immediate environmental effects of use	*none*	*airborne particles unpleasant*	*extensive layer of chemical powder on all surfaces in the area*

➡ **Workbook pages 194/195**

Humidity and evaporation: Review

Speaking

1   Look back at Unit 8.

   Student A: Check that you understand the process of evaporation.

   Student B: Review the idea of humidity.

2   In pairs, explain the concepts to each other.

Listening

You are going to listen to a teacher describing the cooling systems of some traditional Middle Eastern buildings to some students.

1   Look at the photos. Discuss with a partner exactly what they show.

2   🎧 Listen to the first part of the recording and then discuss the questions below with your partner.

   a Why is the air humid?

   b Why does the wind keep us cool?

   c Think of ways in which people could have created systems for cool interior environments before electricity or petrol engines.

3   🎧 Listen to the next part of the recording.

   Label the sketch: house, wind tower, tunnel, air intake.

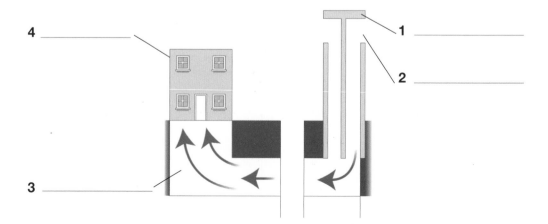

4 🎧 **Listen again. Write short notes to explain how the system works.**

5 **Compare notes with your partner and make any changes as required.**

6 🎧 **Listen to the last part of the recording.**

a Draw a sketch of the system being described.

b Look at this part of the description of the system. Can you remember/guess the missing words?

> Now what happens in this case is that the wind flows over the top of the dome, and as it does so it creates low p_____ , like the airflow over the top of an aeroplane wing. And because of the lower pressure over the roof, the w_____
> a_____ inside the house is pulled upwards and out of those holes. New
> 5   air flows into the bottom of the house, and you get a constant a_____
> through the interior …
>
> … Well, that house also has a big tank, a r_____ , of water under part of the main floor to help keep the a_____ t_____ low. As with the water in the tunnel in the case of this house, e_____ keeps the
> 10   air temperature down.

c 🎧 Listen again and complete the description.

d Which A/C system is which in the photos in Exercise 1?

Speaking and pronunciation

1 🎧 **All of these words relate to air conditioning. How many syllables do they have? Listen and circle the odd one out.**

a environment	(average)	conditioning	contaminant
b system	pressure	falls	intake
c distribution	reservoir	temperature	ventilate
d rise	ducting	moist	cloud
e volume	cool	mixture	moisture
f exhausted	contracting	compression	constant
g humidity	saturate	evaporate	equivalent

2 **Work with a partner. Ask and answer questions about the meaning of these words, for example:**

What is 'constant'?

Constant is an adjective. It means always the same, for example, 'a constant temperature'.

➡ **Workbook pages 196/197**

Lesson 4
Air pressure: Review

Speaking and vocabulary

Work in pairs.

1 **How do you say these numbers?**

⅓   7,450   3.14   4½   159   1,450,000   ± 0.3   7.08   ⁹⁄₁   ⅙   308

2 **Student A: look at list A; Student B: look at list B. Check that you understand the items in your list.**

A	pressurised cabin	climb rate	fuselage structure	outward forces
B	cruising altitude	sea level	cabin pressure	flight engineer

3 **Explain your items to your partner.**

Reading

You are going to read an information leaflet about airline cabin pressure.

1 **Find and circle the following numbers in the text. What do the numbers refer to?**

777   11,000   43,000   7,200   40,000

2 **Expand the prompts below into full questions.**

3 **Find an answer to each question in the text and then ask and answer the questions with a partner.**

Why / important / pressurise / aircraft cabins?

_____

How / aircraft / pressure systems / control?

_____

At what altitude / necessary / pressurise / cabin?

_____

Who / responsible / controlling / cabin pressure?

_____

4 **Find an expression in the text that means:**

a the relationship between the pressure inside and outside the plane.

b the rate of change of the pressure inside compared with the pressure outside the plane.

c the flight crew's control of changes in cabin pressure.

Smaller aircraft, which do most of their flying below 10,000 feet, are generally unpressurised; aircraft which normally operate above 10,000 feet have pressurised cabins.

Today's more modern systems are electronic and are automatically controlled. In aircraft with manual pressurisation systems, the flight engineer's job of controlling cabin
5   pressure is sometimes referred to as "flying the cabin", while the pilots fly the aircraft. This refers to the fact that the cabin pressure or effective altitude inside the aircraft is different from the real altitude of the aircraft, even though as the airplane climbs higher, the altitude in the cabin also rises. There is a relationship between the altitude inside the aircraft and the altitude outside the aircraft, known as the "pressure differential."

10   To give you some idea of cabin altitudes vs. aircraft altitudes, a B-777 cruising at 43,000 will have a cabin altitude of around 7,200 feet. The cabin can maintain sea level pressure up to around 11,000 feet before its maximum pressure differential is reached. At that time, the cabin altitude will start going up on a "climb schedule" which is related to the aircraft's actual climb rate as well as the planned cruising altitude.

15   A common question is why we cannot just fly with a cabin pressure of sea level all the time. Pressurising an aircraft exerts strong outward forces against the fuselage structure. To fly with a cabin at sea-level pressure at normal cruising altitudes of 30,000 to 40,000 feet or so would require so much reinforcement in the aircraft fuselage that it would make the aircraft too heavy to fly.

5   What does this graph show?

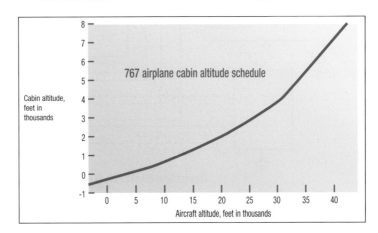

6   Transfer the data from the graph to the table below.

effective altitude	actual altitude
6,000 ft	
	32,000 ft
3,500 ft	
	20,000 ft
1,000 ft	
	sea level
500 ft below sea level approx.	

➡ **Workbook pages 198/199**

Speaking

You are going to read a section of a maintenance manual on electrical systems.

Go to Unit 9 and find information about AC and DC electrical distribution in an aircraft.

Reading

1   **Look at the diagram below. How many of the components in this system can you identify?**

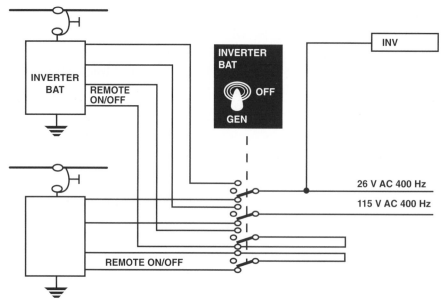

Inverter System Schematic

2   **Use the chart of electrical symbols at the back of the book to find as many of them as you can.**

   a   How many of these abbreviations for electrical systems do you know the meaning of?

   Hz   CB   BAT   DC   GEN   CWS   AC   INV   V

   b   Quickly find and circle all the examples of them in the text below.

   c   Look again in the text and find the full versions of the abbreviations.

AC SYSTEM – DESCRIPTION

The alternating current generation system, which is used to power the attitude indicator, compass and avionics equipment, consists of two identical transistorised static inverters and an inverter selector switch. The system is protected by circuit breakers located in the front cockpit

5 and is monitored by the Central Warning System (CWS). The static inverters, located in the battery compartment, convert 28V DC electrical power to 115 V AC and 26 V AC, 400 Hz. Each inverter is capable of supplying the total load requirements of the aircraft AC systems. Only one inverter is selected on-line at a time; the remaining inverter is used as a standby source of power.

10 The inverters, labelled INVERTER BAT and INVERTER GEN, receive their 28 V DC input separately from the battery busbar and the generator busbar. This prevents complete loss of aircraft AC power in the event of a single DC busbar failure. A four-pole, three-position switch is used for inverter selection. The switch is located in the front cockpit, LH console, and is labelled INVERTER, with marked positions BAT-OFF-GEN. The CWS monitors the 26 V AC

15 output of the on-line inverter, and the letters INV are shown when the voltage falls below 13 volts AC. A 25 amp circuit breaker is located in each of the CB panels. The circuit breakers are labelled INV BAT CB and INV GEN CB.

AC SYSTEM – OPERATION

Both inverters are supplied with a positive DC input when power is connected to the direct

20 current distribution system. When the INVERTER switch is selected to the mid (OFF) position, the inverters are prevented from converting the DC input to AC by internal inhibiting circuits which are controlled by a remote on/off loop circuit routed through the INVERTER switch. Selecting the switch to BAT or GEN enables the selected inverter and connects the nominal outputs of 115 V AC and 26 V AC, 400 Hz, to the aircraft AC distribution system.

---

3 Which of the two sentences in each pair is true, according to the text?

Paragraph 1

a **i** The CWS monitors this system.

  ii This system monitors the CWS.

b **i** The AC system can be supplied by each inverter.

  ii Each inverter can be supplied by the AC system.

Paragraph 2

c **i** Loss of AC power prevents separate 28 V supply to INV BAT/INV GEN.

  ii The separate 28 V supply to INV BAT/INV GEN prevents loss of AC power.

d **i** Inverters are selected using a switch.

  ii A switch is selected using the inverters.

Paragraph 3

e **i** The inverters can be switched off by inhibiting circuits.

  ii The inhibiting circuits can be switched off by the inverters.

f **i** The BAT or GEN position activates the AC supply to the system.

  ii The AC supply activates the BAT or GEN selector.

⇨**Workbook pages 200–202**

Electrical systems: Review II

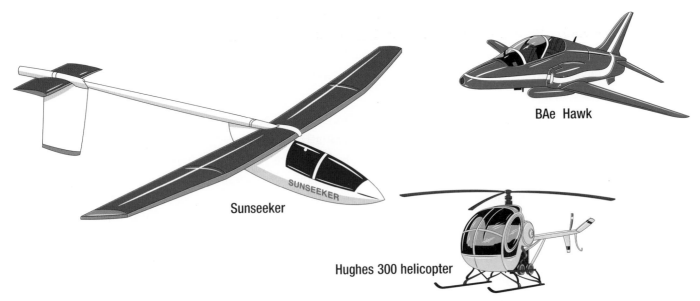

BAe Hawk

Sunseeker

Hughes 300 helicopter

Speaking

Discuss this question with a partner: If electric engines could be developed for aircraft, they would have a lot of advantages. Can you think what these advantages might be?

Listening

1 🎧 Listen to the first part of a talk, which refers to the three photos above. List the advantages of electric aero engines.

2 Check with a partner. How many advantages did you note? Were any of your ideas from Speaking mentioned?

3 🎧 Listen to the next part of the talk and label the diagram using these items: magnets, shaft, armature, electromagnets, rotor.

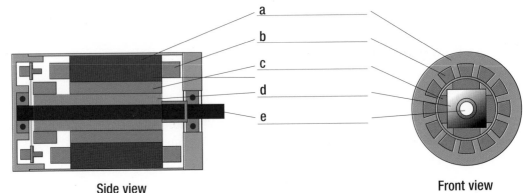

a

b

c

d

e

Side view

Front view

Reading and speaking

The instructor went on to talk about other aspects of electric engines. Read a student's notes from the rest of the talk.

1   What was the topic of the rest of the talk: 'Further advantages of BDLC', 'Disadvantages of electric engines' or 'The Sunseeker solar plane'?

> Disadvs.
>
> Improved motors, e.g., BLDC, not whole story. Still prob of the power supply:
>
> 1   Even modern bats. can't convert chem. energy into elec. energy effectively, c.f., kerosene.
>
> 2   Bats heavy. Sunseeker can carry pilot cos it gets power from solar cells – convert sunlight into elec. – so no need to carry heavy bat.
>
> 3   Plane doesn't get lighter as it flies, unlike w. kerosene. Only way to get round prob. wd be throw bats out of plane when used up. Not prac.
>
> fig 2 on h/o is chart of energy prod./kg.

2   Look at the table. What did the lecturer say about different types of electric aero engine batteries?

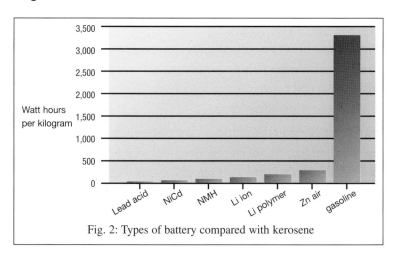

Fig. 2: Types of battery compared with kerosene

Writing and language

Comparison

1   Find and underline these expressions in the tapescripts.

> far more   as ... as   much more ... than

2   Expand the information in the student's notes into full sentences. Try to include some comparative expressions.

⇨ **Workbook pages 202/203**

Instruments: Review I

Speaking I

Look at the diagram and the labelling, and discuss these questions with a partner.

a Which of the 'basic six' is this instrument?

b How do you think it works?

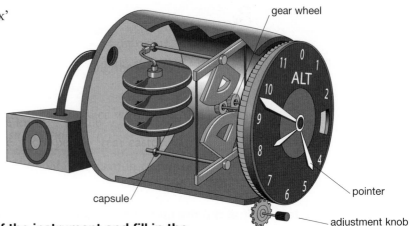

gear wheel

ALT

capsule

pointer

adjustment knob

Reading

1 **Read the description of the instrument and fill in the gaps with the labels and information from the diagram above.**

Inside the altimeter there are one or more _____. These are sealed, airtight, flexible containers capable of expansion and contraction. They are stacked with one end of the capsule fastened so that it cannot move. The other end extends or retracts as the pressure inside the instrument changes.

5 The pressure of the atmosphere outside the plane is transmitted via vents open to the outside air at the rear of the fuselage. As the aeroplane climbs and external air pressure falls, the fixed quantity of air in the capsule will expand and cause the capsule to extend. When the aeroplane descends and the static pressure increases, the capsule will contract.

10 The moving end of the capsule is attached to the altitude display via a shaft and _____. These are usually made of brass.

The display has three _____ which give the aircraft's altitude in tens, hundreds and thousands of feet. NB: In British and US aircraft, it is still standard to use feet, not meters.

15 Because the pressure inside the aircraft cabin will be different from the external pressure, the body of the instrument is airtight, except for the opening to the

_____.

Since barometric pressure changes at sea or runway level, the pilot is able to adjust the instrument to change the level at which the instrument reads zero feet. This is done

20 using the _____. If there has been any change in the weather, the pilot will need to set the pointers to zero before he takes off.

2 Match an item from each column to make phrases.

fastened	inside	the instrument
the pressure	to	the rear of the fuselage
the atmosphere	outside	the body of the instrument
the ports are	to	the altitude display
attached	to	zero
set a display	at	the plane

3 Complete the sentences with one of the expressions from the centre column in Exercise 2.

a Before beginning the test, set the throttle valve _____ zero.

b The oxygen supply is fastened with a bolted clamp strap _____ the seat.

c On a 737, the ELT is located _____ the rear of the fuselage in front of the tail.

d If the pressure _____ the balloon increases, the balloon will rise.

e Some instruments can be affected by weather conditions _____ the aircraft.

f Hoses and tubes are often attached _____ other tubes using bayonet connectors.

Speaking II

Work with a partner. Look at this diagram of another flight instrument, the airspeed indicator. Can you explain how it works?

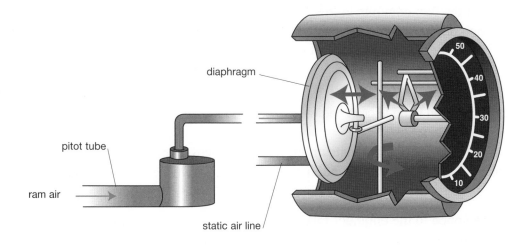

diaphragm

pitot tube

ram air

static air line

50
40
30
20
10

⇨**Workbook pages 204/205**

Speaking

1   Discuss the following questions with a partner.

    a   What are the 'basic six' flight instruments?

    b   What does each one do?

    c   Can you remember their relative positions on the cockpit display?

2   **Check your answers by looking at Unit 10, Lesson 2.**

3   **Which of the two cockpits in the photos seems to be more traditional? How is the other one different?**

Listening

You are going to listen to an aircraft engineer describing modern glass cockpits.

1   Look at the glass cockpit instrument display above. It contains all of the basic six displays. Can you see where each one is?

2   🎧 Listen once to the engineer talking on a radio programme and answer these questions:

    a   Why is it called a 'glass' cockpit?

    _____

    b   What additional information is given in the display apart from the basic six flight instruments?

    _____

3 🎧 Listen again and label the basic six flight instruments.

Writing and speaking

1 Work with a partner. Read these notes about the glass cockpit.

> Glass cockpit
>
> 'Glass' refers to the PFD – Primary Flight ... (?)   basic six info replaced
> by completely new info.   All info on one screen to make landing easier
> Instr.:   centre – att. indic / airspeed indic
>   L/H side – gyro compass / vertical speed indic
>   R/H side – altitude

2 Some of the information is incorrect. Correct the information according to the information given by the engineer on the radio programme.

3 🎧 Listen to the second part of the interview and make your own notes on:

a the Instrument Landing System.

b the Navigation Display.

4 Check your notes with a partner.

5 🎧 Listen again to the recording and follow the tapescript to check your work.

6 Practise repeating the recorded voices phrase by phrase. Copy the voices as closely as you can.

Language

If

What would happen if the pilot did not have the information given to him by his instruments?

1 Write sentences about the six basic flight instruments and the navigation, communication and radar systems. For example:

If he didn't have an altimeter, he wouldn't know how high he was flying.

2 Compare with a partner.

⇨ Workbook pages 206/207

Vocabulary

Troubleshooting flow charts help technical personnel to trace faults in a clear and systematic way.

1   Look at the words in the box below and divide them into the following catergories.

- equipment parts and components
- problems

> fuse   thermostat   socket   overheat   PCB   fan   sensor   terminal   pressure switch
> lead   bad connection   gas valve   fault   sparking   blockage   burner

2   What piece of equipment do you think they are all connected with?

Reading

The maintenance technician needs to go through certain checks.

1   Complete these questions with *is / is there / can*.

a   _____ the fan turning?

b   _____ mains on neon '13' illuminated?

c   _____ a live supply to both terminals of overheat 'stat?

d   _____ the overheat 'stat be reset when the system is cold?

e   _____ a supply to fan connector on PCB?

f   _____ the fuse blown?

g   _____ the ignition electrode sparking?

h   _____ the boiler casing correctly fitted?

i   _____ a live supply on the lead to pressure switch?

j   _____ the pilot injector blocked?

2   Work in pairs, A and B, with the flow chart on page 253. It shows how to identify and check faults in a domestic boiler.

a   Read through the chart quickly.

b   Student A: Decide on a fault.
   Student B: Use the chart to ask Student A questions at each stage.
   Student A: Answer, for example, *Yes, it is / No it can't*, etc., and B will ask you the next
   question until the flow chart gives you a final decision.

3   Check that you have both reached the same place on the chart.

Writing

Look at these instructions for checking a small piece of electrical equipment.

> If the equipment doesn't work at all, check for power at the supply. If there is no supply detected, then switch on power, and if this doesn't work, check the fuse or circuit breaker and replace or reset. If the appliance still doesn't work, carefully inspect the cable for damage and replace if necessary. Switch on the equipment again and if there is still no result, call the service technician.
>
> WARNING
>
> If there are signs of burning around the fuse or circuit breaker, do not replace or reset until the cable has been checked.

1　**Write the questions which need to be asked at each stage of the troubleshooting process.**

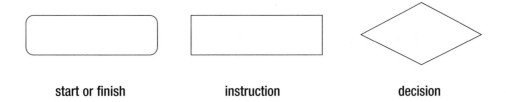

| start or finish | instruction | decision |

2　**It is easier to understand a flow chart if different shapes are used for each of the different kinds of box. Which of the boxes should be used for:**

　a　engine starts Y/N?

　b　check all contacts and soldered connections?

　c　replace PCB

　d　contact the component manufacturer with full test results

　e　run the motor up to 2,500 RPM?

3　**Draw and complete the fault-finding flow chart for the equipment.**

4　**Design and write a troubleshooting flow chart for a mobile phone that will not operate. The first box should be:**

> Mobile does not work

⇨**Workbook pages 207–209**

Speaking

Look at these drawings and discuss these questions with a partner.

1   **What do the drawings show?**

2   **Why are the three jacking points in these positions?**

3   **Discuss how, working with a colleague, you would lift the aircraft off the ground as shown in the drawings. Use this language.**

> We'd need to …   I'd have to be careful to …   I'd …   Then …   After that …

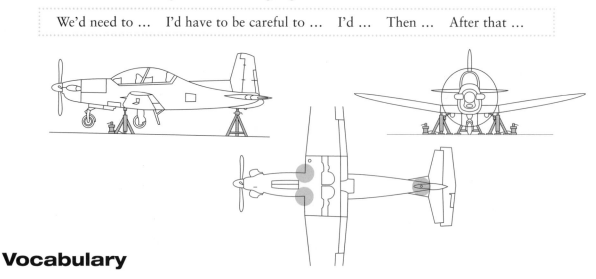

Vocabulary

Write one of these verbs in the space in each sentence.

> raise   lower   add   extend   adjust   position   install   open   continue   insert

1   _____ the aircraft tail to the required height.

2   _____ the central pillar of the jack.

3   _____ the aircraft tail down onto the stand.

4   _____ the jack positions as necessary.

5   For added safety, _____ weight to the aircraft tail.

6   _____ main jack hydraulic valves.

7   _____ raising the aircraft slowly and evenly.

8   _____ and secure a jacking adapter on top of each of the two hydraulic jacks.

9   _____ the jacks under the main jacking points.

10 _____ the locking pin through the centre tube and pillar.

Reading

Look at the instructions for jacking.

1 **Find and underline all the phrases and sentences in Vocabulary above.**

2 **The jacking instructions are not in the correct order. Read them again and discuss the correct order with a partner.**

— **A** NOTE: In the following step, two persons are required, one to raise and lower the aircraft tail and the other to adjust the tail stand. Raise the aircraft tail to the required height, withdraw the tail stand locking pin and extend the central pillar of the jack to the approximate new height. Install the stand locking pin in the nearest locking hole and lower the aircraft tail down onto the stand ensuring that the stand legs are firmly in contact with the ground.

— **B** Slowly and evenly raise the aircraft until its leg struts are fully extended and the main wheels are just clear of the ground.

— **C** Adjust the tail height again if required.

1 **D** Ensure that there are no obstructions around or above the aircraft.

— **E** Ensure that the jack hydraulic valves are closed. Raise the jacks until the adapters contact the jacking points, adjusting the jack positions as necessary to align the adapters and jacking points.

10 **F** For added safety, add weight to the aircraft tail by suspending ballast of approximately 40 kg (88 lb) from the ballast attachment point.

— **G** Slowly open main jack hydraulic valves and gently lower the aircraft onto the jack locking collars. Close the hydraulic valves. The jacks are now hydraulically and mechanically locked.

— **H** Position the tail stand under the tail jacking point. Raise the stand central pillar to the approximate height of the jacking point and insert the locking pin through the centre tube and pillar. Adjust the tail stand threaded ball-end until it contacts the jacking point, ensuring that all three legs of the stand are firmly in contact with the ground.

— **I** Continue raising the aircraft slowly and evenly at the main jacking points until locking collars can be installed. Raise the aircraft tail to the required height (as detailed in Step 6), finally adjusting it to the required level by use of the threaded ball-end.

— **J** Install and secure a jacking adapter on top of each of the two hydraulic jacks and position them under the main jacking points.

Listening

🎧**Listen to an experienced technician showing a trainee the jacking procedure, and check the order of the instructions.**

➡️**Workbook pages 209/210**

Unit 1 Lesson 6: Language and speaking

Student A

Information sheet for Boeing 747 passenger airliner

length	70.6 m
width (wingspan)	64.4 m
height	19.4 m
cabin length	60 m
crew	33
passengers	524
max. take-off weight	396,890 kg
payload	112,760 kg
power plant	4 x Pratt and Witney 40,000 hp 4062 engines
maximum speed	968 km/h
ceiling	9,600 km
range	13,716 m

Unit 1 Lesson 6: Language and speaking

Student B

Information sheet for Lockheed Hercules C-130 military transport plane

length	29.8 m
width (wingspan)	40.4 m
height	11.6 m
cabin length	16.7 m
crew	4–6
passengers	92
max. take-off weight	70,300 kg
payload	20,000 kg
power plant	4 x Allison T56 4,300 hp engines
maximum speed	610 km/h
ceiling	6,800 km
range	10,000 m

Unit 3 Lesson 5: Writing and speaking

Student A

Monocoque airframe

The word monocoque is, in fact, a French term meaning 'single shell'. In this airframe design, the 'shell' – the skin of the aircraft – becomes a structural element, and bears much of the shear and bending forces on the airframe during flight. The great advantage of this system is that a light skin, e.g., aluminium, which has a small volume can handle a high load. The load is transferred from a heavy component – an internal frame – to a light one – the skin – so weight is saved. Another advantage is that there is much more space inside the aircraft.

A common example of monocoque construction is in an ordinary cardboard box, where two sheets of brown paper cover a light supporting structure. This 'honeycomb' idea is now used inside the skin of military aircraft. Modern Formula 1 racing cars are another example of the use of a load-bearing skin over a simple chassis.

Materials such as carbon fibre are especially suitable for monocoque construction. Carbon fibre can be designed to be strong in some directions and simultaneously flexible in others. It is very strong, very light, and also easier to work with than steel or aluminium.

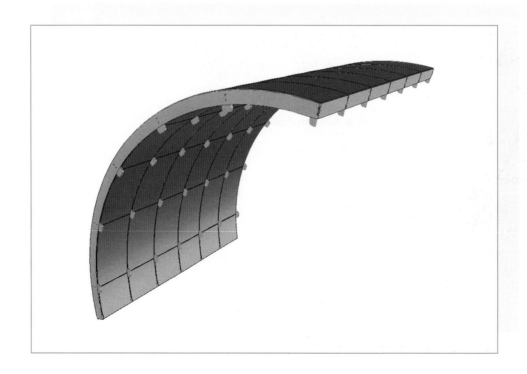

Unit 3 Lesson 5: Writing and speaking

Student B

Trussed airframe

A traditional 'trussed' airframe construction uses a load-bearing internal frame with a separate skin covering it. The internal frame is usually made of wood or steel tubes, and the skin is made from, e.g., sheet metal, plywood – or simply fabric. Although this design is strong in all directions, it is heavy, and the space inside the aircraft is taken up by the frame. Also, the frame is often easy to build by hand but is not suitable for mass-production.

This type of airframe is now old-fashioned, and it was not widely used after the 1930s. This is because. in the late 1920s. the very light and strong metal, aluminium, which had been expensive before, became cheap. Different variants of all-aluminium airframe construction were soon trialled: firstly, a large aluminium frame was used with a thin, riveted aluminium skin; secondly, a small frame with a thick skin was tried; and thirdly, the combination of a light frame with a stressed skin. This last option gave the best combination of strength and lightness, and is still the basic design for general-purpose aircraft.

Unit 4 Lesson 5: Listening

Process of an aircraft landing

1 _____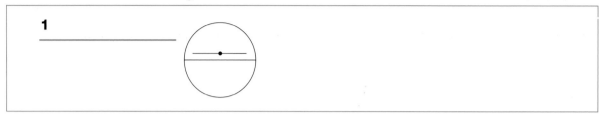

2 _____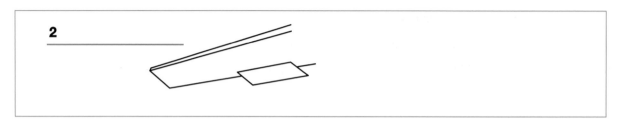

3 _____

ALTITUDE
2500 ft

4 _____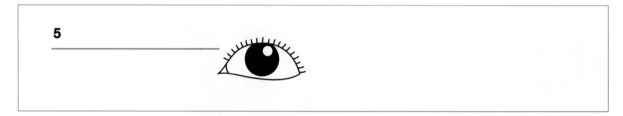

5 _____

6 _____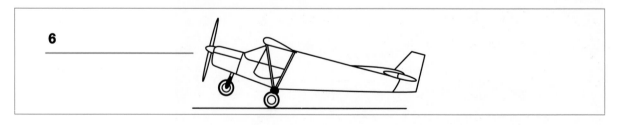

7 _____

Take-off **Additional material**

Unit 5 Lesson 3: Speaking and writing
Skeleton notes

Complete the notes using some of the words from Speaking and writing Exercise 1.

Example:

a Have you ever watched an ice-skater on TV?

- Skater uses arms and legs to change speed of _rotation_ .

- Can _increase_ velocity without using extra power/decrease velocity without using brakes?

b The answer is momentum.

- No change in _____ without external force.

- Applies to all situations: straight line/rotary/angular – momentum is _____.

c Think of the skater in mathematical terms for a moment.

- hands and leg (M) rotate (V) – travel specific _____ round circumference of _____ in specific time.

- Radius of circle r decreases if arms pulled in towards body as _____ travelled by hands is smaller.

- V _____ as overall _____ must remain same, i.e., _____ spins faster.

- V stable state due to unchanging momentum (similar to stability in gyroscope and _____).

d This fact was used to provide the answer to an old and difficult problem.

- Early in-line piston engines: too much linear _____ strain on the airframe, heavy flywheel needed to provide angular _____.

- _____ overcome by _____ engine: combustion cylinders arranged in circle so that _____ evenly distributed (like ice-skater).

- Engine more _____, ran smoothly, less _____ on airframe.

Unit 3 Lesson 5: Writing and speaking

Student A

Text A: Advantages of the rotary engine

One solution to the problem of vibration was a type of rotary engine. The crankshaft was stationary (did not move) and the cylinder block acted as the flywheel in this engine. This provided the advantages of the circular arrangement of cylinders without the extra weight. The propeller was attached to the cylinder block. Another advantage of this type of engine was that it was easier to start by hand because the momentum of the heavy cylinder block kept it rotating for long enough to start the firing cycle of the engine. Also, like the radial engine, it could be air-cooled because the cylinders were all in the same plane behind the propeller. There was no need for cooling fluid or a radiator.

Advantages

Take-off **Additional material**

Unit 3 Lesson 5: Writing and speaking

Student B

Text B: Disadvantages of the rotary engine

The rotary engine had some major disadvantages. It could only operate at full power, which meant that it used a lot of oil and fuel. It also made the plane more difficult to manoeuvre because the strong gyroscopic effect tended to resist the pilot's changes of direction. This was because although the overall weight of the power plant was reduced, the rotating weight was increased. For example, the Sopwith Camel, a famous British fighter plane of World War I which was fitted with this kind of engine, was easy to turn to the right, but difficult to turn to the left. Nevertheless, because of its excellent power-to-weight ratio, large numbers of this type of engine were produced between 1910 and 1918, before it was replaced by better technology.

Radar diagrams

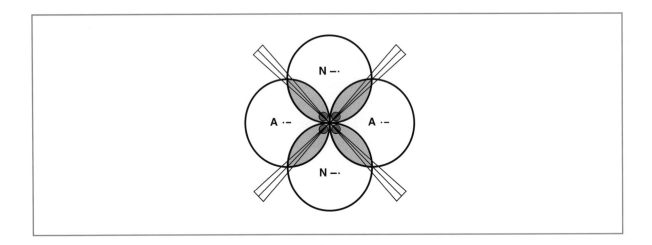

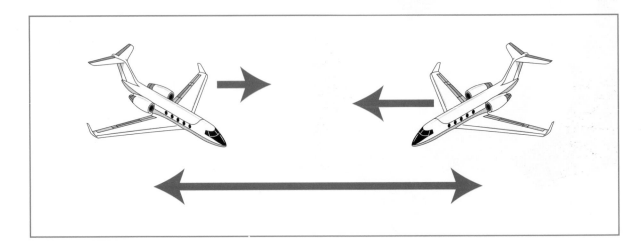

Take-off Additional material

Unit 11 Lesson 1: Speaking

Forms

Form 1

DEPARTMENT OF TRANSPORTATION FEDERAL AVIATION ADMINISTRATION MALFUNCTION OR DEFECT REPORT		OPER Control No.		8. Comments (Describe the malfunction or defect and the circumstances under which it occurred. State probable cause and recommendations to prevent recurrence.)	DISTRICT OFFICE	OPERATOR DESIGNATION
		ATA Code				
		1. A/C Reg. No.	N.		OTHER	
Enter pertinent data	MANUFACTURER	MODEL/SERIES	SERIAL NUMBER			
2. AIRCRAFT					COMMUTER	
3. POWERPLANT						
4. PROPELLER					FAA	
5. SPECIFIC PART (of component) CAUSING TROUBLE					MFG	( )
Part Name	MFG. Model or Part No.	Serial No.	Part/Defect Location			
					AIR TAXI	
6. APPLIANCE/COMPONENT (Assembly that includes part)					MECH.	TELEPHONE NUMBER (
Comp/Appl Name	Manufacturer	Model or Part No.	Serial Number	Optional Information:	OPER.	
Part 11	Part TSO	Part Condition	7. Date Sub.	Check a box below, if this report is related to an aircraft: ☐ Accident: Date _____ ☐ Incident: Date _____	REP. STA.	SUBMITTED BY:

Form 2

US Department of Transportation FEDERAL AVIATION ADMINISTRATION	SUSPECTED UNAPPROVED PARTS REPORT	
Refer to page 2 for instructions on how to complete the form.		
1. Date the Part was Discovered:		2. Part Name:
3. Part Number:		4. Part Serial Number:
5. Quantity:	6. Assembly Name: Assembly Number:	7. Aircraft Make & Model:
8. Name, Address, and Description of the Company or Person Who Supplied or Repaired the Part:		
Name:	Street Address:	
City:	State:	ZIP Code:
Country:	Phone Number:	

Forms

Form 3

DEPARTMENT OF TRANSPORTATION FEDERAL AVIATION ADMINISTRATION MALFUNCTION OR DEFECT REPORT	Form No: AWSD Form 6A AMDT No: 0 Issue Date: 17/11/00 Page No Page 1 of 5	PRELIMINARY INSPECTION REPORT FOR ISSUE/RENEWAL OF CERTIFICATE OF AIRWORTHINESS Level 1

IRISH AVIATION AUTHORITY
Airworthiness Standards Department
Aviation House, Hawkins Street, Dublin 2

Preliminary Inspection Report For Issue/Renewal of Certificate of Airworthiness

Completed form including the Airworthiness Directives Listing and Equipment List to be forwarded to above address.

Note: *Where an item is not applicable, or appropriate, the letters 'NA' should be entered.*

1. Registration: _____ 2. Aircraft Type: _____ 3. Serial No.: _____

4.	Since Manufacture	Since last C of A	Last Calendar Year
Aircraft Hours			
Aircraft Cycles/Landing			

Form 4

US Department of Transportation FEDERAL AVIATION ADMINISTRATION	**MAJOR REPAIR AND ALTERATION** **(Airframe, Powerplant, Propeller, or Appliance)**	Form Approved OMB No. 2120-0020 11/30/2007
		For FAA Use Only
		Office Identification

INSTRUCTIONS: Print or type all entries. See FAR 43.9, FAR 43 Appendix B, and AC 43.9-1 (or subsequent revision thereof) for instructions and disposition of this form. This report is required by law (49 U.S.C. 1421). Failure to report can result in civil penalty not to exceed $1,000 for each such violation (Section 901 Federal Aviation Act of 1958).

1. Aircraft	Make	Model
	Serial No.	Nationality and Registration Mark
2. Owner	Name (As shown on registration certificate)	Address (As shown on registration certificate)

3. For FAA Use Only

4. Unit Identification				5. Type	
Unit	Make	Model	Serial No.	Repair	Alteration
AIRFRAME	----------------------(As described in Item 1 above)------------------------				
POWERPLANT					
PROPELLER					

Unit 12 Lesson 1: Reading

Fault-finding flow chart

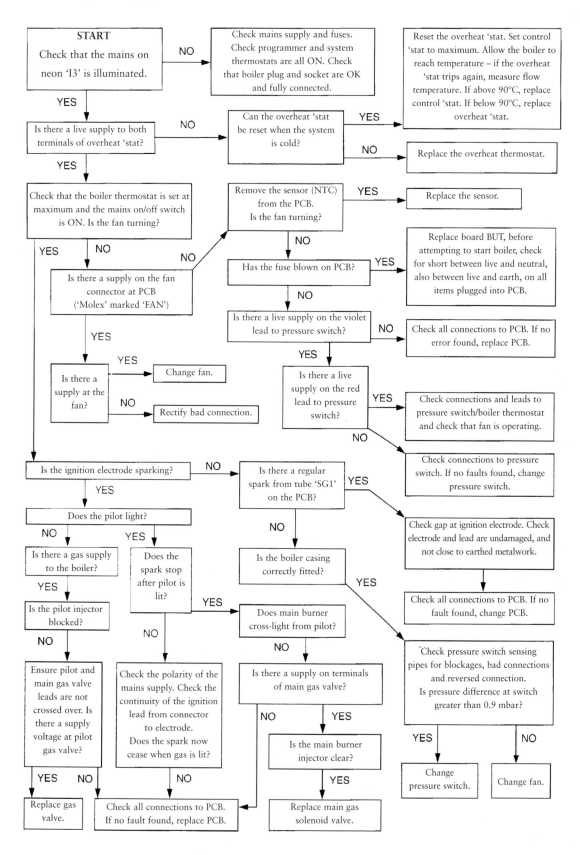

Glossary

AC (alternating current) (n) electricity which flows through a circuit, first in one direction and then in the opposite direction. The number of changes of direction per second (the frequency) is measured in units called Hertz.

accumulator (n) a device used to store hydraulic fluid under pressure. When a valve is opened, hydraulic fluid is forced out of the accumulator by compressed gas and can be used to operate equipment.

actuator (n) a device which changes hydraulic or electrical energy into mechanical energy in order to do work. Valves can be actuated by electric motors and air brakes by hydraulic pistons.

aerospace industry (n) all the companies that design, manufacture and maintain aircraft, rockets and spacecraft.

aileron (n) a hinged part of the aircraft wing which can be raised or lowered to control the turning and rolling movement of the plane.

air brake (n) a hinged surface, lowered from the aircraft when it is landing, which helps to slow the plane by increasing air resistance.

air duct (n) a channel or tube which carries air. Clean air is circulated through an aircraft by means of air ducts.

air traffic controller (ATCO) someone who makes sure that aircraft in the air are safely separated from each other, both vertically and horizontally.

airframe (n) the main body of the aircraft consisting of the fuselage, wings, tail and skin, but without the engines, instruments or any other internal equipment.

airworthy (adj) in a good enough condition to fly safely and efficiently.

alloy (n) a metal which is made by mixing two or more liquid (molten) metals together. For example, brass is an alloy of copper and zinc. Solder is an alloy of lead and tin.

alternator (n) a type of generator using rotary mechanical motion of an engine to produce an alternating current

altimeter (n) an instrument which measures the vertical distance above sea level (the altitude) of an aircraft by means of air pressure changes or radio signals.

ammeter (n) an instrument which measures the quantity of electric current (the number of *amperes* or *amps*) flowing in a circuit.

ampere (amp) (n) the basic unit of the quantity of electric current. Very small currents are measured in **milliamps** (thousandths of a milliamp).

amplifier (n) an electronic device which increases the strength of an electrical signal.

analogue multimeter (n) an instrument which can be set to measure different electrical quantities, such as current, voltage and resistance. The position of the pointer on the dial of the analogue multimeter varies directly with the quantity it is measuring.

antenna (aerial) (n) a device connected to a radio transmitter or receiver which is used to send or receive signals, usually mounted on the outside of the plane.

armature (n) part of the rotating shaft of an electric motor or generator which has a coil or coils of wire wound on it.

artificial horizon (n) an instrument which shows a pilot the position of his aircraft relative to the real horizon, without needing to see outside the plane. It consists of a horizontal line which rotates clockwise or anticlockwise as the plane rolls, climbs or descends.

assembly (n) a group of parts connected together (assembled) to make another part, such as the gearbox of a car or the tail section of an aircraft.

ATC (n) an abbreviation of Air Traffic Control.

avionics (n) the electronic communication, navigation and control equipment of an aircraft.

B

backup system (n) equipment which is only used when the main equipment fails. Aircraft may have one, two or even three backup systems for particular pieces of equipment. For example, if a generator fails, power can be supplied by a backup battery instead.

battery (n) a device used for storing direct current electricity.

bearings (n) these support a moving part in the correct position but allow it to move freely. For example, electric motors have bearings at each end of the shaft, which keep it in position but allow it to rotate easily.

blind rivet (n) A type of *rivet* which can be fixed from one side of the workpiece only. Blind rivets are used when it is impossible to get to the other side of the workpiece, or when there is only one person to do the work.

boost (v) to increase the strength or the supply of something above the normal amount.

booster pump (n) a pump which can supply more fluid or higher pressure than normal.

bracket (n) a piece of metal, usually in the shape of an L or a triangle, which is used to support and hold individual components or complete items of equipment.

brake (v & n) (v) regular verb which means to slow down or stop a moving vehicle; **(n)** a device used to slow down or stop a moving vehicle.

bridge crane (n) a type of overhead mobile crane used to lift heavy loads in hangars, workshops and factories. The lifting equipment is suspended from a high steel beam (the bridge). It can move along the beam, which may also be able to move within the workplace.

brittleness (n) a property of materials which means they break into pieces easily when hit. Biscuits, glass and cast iron are examples of brittle materials.

bulkhead (n) a wall which divides one part of the aircraft from another in order to provide extra structural strength or safety. Small single-engine planes often have a fire-proof bulkhead between the engine and the cockpit.

burner (n) a device for controlled burning of fuel. Some jet fighter aircraft have afterburners, which inject fuel into the hot exhaust gas, providing sudden extra power.

burrs (n) small, sharp pieces of material forming the rough uneven edge of holes, sheets or workpieces which have not been properly finished and made smooth. Burrs can cause injury and damage.

busbar (n) large conductor with a high current capacity, used to connect specific power supply to a number of pieces of equipment.

C

camshaft (n) an engine shaft which has irregular-shaped sections (cams) mounted on it which open and close valves as the shaft rotates.

capacitance (n) the ability of a component or system to store a DC static charge. Capacitance can be measured with a capacitance meter in units called Farads.

capacitor (n) an electronic component which stores a DC static electric charge and will conduct AC current.

centre of gravity (n) the point on an aircraft where it is balanced in all directions. On modern aircraft, it is between the nose wheel and the wings. On older aircraft (taildraggers), it was between the wings and the tail.

centrifugal (adj) moving away from the centre of something that is rotating.

chassis (n) structure in the form of a metal box or frame on which components are mounted. In the past, cars were built on a chassis. Electronic equipment is often built on a chassis.

check valve (n) a valve which allows a liquid or gas to flow in one direction only. The simplest types contain a ball and spring or a hinged flap.

circuit board (n) a sheet of insulated material containing a number of connected electronic components. Sometimes, the board will have a circuit already printed on it, ready to receive the components.

circuit diagram (n) a drawing consisting of lines and component symbols which show how an electrical or electronic device works and how its components are connected to each other.

circuit-breaker (n) an automatic or manual

emergency switch which disconnects the electrical power supply before equipment becomes overloaded and damaged or dangerous.

CNC (computerised numerical control) machine tools (n) lathes, drills, milling machines, etc., which are controlled by signals from a computer instead of being controlled by the hands of an operator. They can accurately produce standard components very quickly.

cockpit (n) the section of an aircraft which contains the pilot, instruments and flight controls. Training aircraft often have two cockpits.

coefficient of linear expansion (n) a measure of the increase in length of a material when it is heated. Each material has a different C of LE which is used to calculate its expansion at different temperatures. Aluminium expands more than cast iron when heated and has a higher C of LE (aluminium = 0.000022/C0 , cast iron = 0.000011/C0).

combustion chamber (n) an enclosed space inside a *turboprop, turbofan* or jet engine where the fuel is burnt. Burning (combustion) of the fuel creates hot, expanding gases which produces *thrust*.

commutator (n) rotating part at the end of the shaft of a DC electric motor or generator which supplies or collects current.

component (n) a separately made part of a device or piece of equipment.

compression stroke (n) the stage in the operation of the internal combustion engine when the piston moves into the cylinder and compresses the fuel-air mixture prior to combustion.

compressor (n) a pump or machine that increases the pressure of a gas or vapour and decreases its volume.

conductivity (n) (adj. conductive) the property of allowing heat or electricity to pass through a material.

conduit (n) a length of pipe, duct or channel used to isolate and protect smaller pipes or cables.

control column (n) the main manual control used by the pilot to control the pitch and roll of an aircraft.

control surfaces (n) moveable sections of the wings and tail of an aircraft which are under the control of the pilot and enable him to change the position and direction of the aircraft.

corrosion (n) a chemical process which can damage or destroy a metal. It is the result of a chemical reaction between the metal and another material, such as acid or water.

cylinder barrel (n) the hollow inside of a cylinder in which a piston moves.

cylinder block (n) part of an internal combustion engine that contains one or more cylinder barrels.

cylinder head gasket (n) a type of sealing washer which ensures a tight seal between the cylinder head and the cylinder block.

D

decoy (n) a protection device used by fighter aircraft to attract enemy missiles to the wrong place so that they do not hit the fighter.

defogging nozzle (n) the end of a small tube which passes warm, dry air onto the cockpit glass to prevent condensation or fogging in order to maintain clear visibility out of the aircraft.

density (n) (adj. dense) the mass (or weight) of a material contained in a specific volume of space. A cubic centimetre of steel weighs more than a cc of aluminium, and is therefore denser.

digital multimeter (n) an instrument which can be set to measure different electrical quantities, such as current, voltage and resistance. Unlike analogue meters, it has no moving parts, and the results of a measurement are displayed as separate digits.

direct current (DC) (n) an electric current that flows continuously in the same direction from one pole to another.

display panel (n) a separate panel in the *cockpit* which holds flight information instruments and warning lights.

drogue parachute (n) a small parachute which comes out first and helps to pull out the main parachute.

ductility (n) (adj. ductile) the ability of a material to be stretched (drawn) into a thinner and thinner wire. Copper and aluminium are used for electric cable because of their high ductility.

durability (n) (adj. durable) the ability of a material to last for a long time without changing. Glass, gold and stainless steel are all durable materials.

ejection seat (n) a seat in a military aircraft which can be shot quickly out of the aircraft in an emergency, with the crewman still sitting in it.

elasticity (n) (adj. elastic) the ability of a material to regain its original shape after stretching or compression.

electrical arc (n) a spark caused by an electrical discharge between two adjacent terminals or contacts. Arcing in electrical switches is sometimes the cause of fires.

electrical resistivity (resistance) (n) the property of preventing the flow of electricity through a material measured in ohms.

electrolytic capacitor (n) a type of capacitor which consists of electrodes separated by a chemical substance called an electrolyte.

electromagnet (n) a piece of ferrous material which is made magnetic when an electric current is passed through a coil which is wound round it. Relay switches are often operated by electromagnets.

elevators (n) moveable control surfaces, usually mounted on the tail, which are used to vary the pitch by forcing the nose of the aircraft up or down.

engine block (n) the main part of an internal combustion engine, containing the cylinder block.

evaporator (n) a device which enables a liquid to change to a vapour (evaporate) easily.

exhaust (n) the escape of waste gases from the rear of an aircraft or other vehicle.

exhaust duct (n) a type of pipe used to convey *exhaust* gases from the engine to the outside of a vehicle.

external power unit (EPU) (n) equipment which supplies electrical power to an aircraft when it is on the ground and bypasses the aircraft's own generator and battery systems.

fan belt (n) a belt of flexible material which connects the main shaft of an engine with a cooling fan.

filament bulb (n) a traditional type of glass lamp which contains a thin wire which gives off light when a current is passed through it.

filter (n) a device for removing unwanted material from a liquid or gas.

fin (n) the vertical surface on the tail of an aircraft (the vertical stabiliser).

flammable (adj) easy to burn.

flange (n) a protruding edge of a component used to position or connect it.

flashpoint (n) the temperature at which the vapour of a flammable liquid such as kerosene will suddenly start to burn without any external source of ignition.

flexibility (n) the ability to be bent out of position and then back again. The wings of aircraft are designed to move up and down (flex) during flight, reducing stress on the airframe.

flight control systems (n) the electronic, electrical and mechanical equipment and instruments which are used to control the movement of an aircraft in the air.

flight data recorder (n) (black box) a very tough and well-protected container of instruments which continuously records details of a flight so that this information is available in the event of an accident.

flight path (n) the line or course flown by an aircraft through the air.

flywheel (n) a heavy wheel which continues to turn after power is disconnected. Flywheels are used to maintain smooth continuous rotation in engines and other equipment.

foreplane (n) small stabilising wings attached to the nose of jet fighters and missiles.

fuse (n) a protective device consisting of a piece of conducting material which melts and disconnects a circuit when there is too much electric current flowing through it.

fuselage (n) the central body of an aircraft without the wings and the tail.

G

gas compressor (n) a machine designed to increase the pressure of a gas by using cylinders and pistons or a series of powerful fans.

gasket (n) a piece of strong sealing material held tightly between two connected parts to prevent any leaks of liquid or gas.

GPS (global positioning system) (n) A navigation system which uses the position of satellites in orbit around the earth (the globe) to detect the exact position of an object on the surface of the earth.

grommet (n) a piece of flexible material fitted into a hole to prevent damage to pipes or cables by the sharp edges of the hole.

groundloop (n) a sudden accidental turning movement of an aircraft that is moving along the ground.

gyro-compass (n) a direction-finding instrument which uses a gyroscope instead of a magnetic needle to indicate the direction of travel.

gyroscope (n) a rapidly rotating wheel that maintains the position of its axis of rotation even when its container changes direction.

gyro-system (n) any control or instrument system which is based on the use of a gyroscope.

H

hand-propping (n) starting an aircraft engine by quickly pulling a propeller blade downwards by hand.

hangar (n) a large building which is used to store, repair or build aircraft.

I

induction stroke (n) the stage in the operation of an internal combustion engine when the piston moves away from the top of the cylinder and sucks air/fuel mixture into it through a valve.

internal combustion engine (n) any engine that produces power by rapidly burning fuel inside a cylinder containing a moving piston.

isolator (n) a device which is used to cut all mechanical or electrical connections between two parts of a system.

J

jack (n) a device which uses gears, compressed air or hydraulic fluid to produce a mechanical advantage to lift heavy loads.

jet pump (n) a type of fuel pump without moving parts which uses a liquid stream of fuel to suck more liquid into the stream.

joystick (n) the main control column of an aircraft.

L

landing gear (n) the extendable wheel and leg assembly usually mounted beneath the wings which, together with the nosewheel, enables it to land safely.

landing gear leg (n) an extending/retracting strut to which a landing wheel is attached.

lead (pronounced 'leed') screw (n) a rotating shaft with a pitched thread which is attached to a lathe. It enables bolts and screws to be made accurately by controlling the speed of the cutter along the workpiece.

lightness (n) the property of not being heavy. The lightness of aluminium is one of the main reasons for using it in aircraft.

locking pin (n) a small metal rod which is inserted into a folding or extending mechanism (such as a landing gear leg or a jack) to make sure that it doesn't change position suddenly.

longeron (n) an important structural beam which extends along the length of the fuselage.

longitudinal axis (n) the centre line which goes through the length of something. The longitudinal axis of a plane runs through the centre of the aircraft, from nose to tail.

lubricity (n) the property of reducing friction so

that parts can move easily against each other. Some fuels have a high lubricity and so reduce engine wear.

lug (n) a small piece of a component or device which projects out of the main body and is used to attach it to something else.

M

malleability (n) the property of being beaten into a thin sheet or a different shape without breaking. Gold and lead are both very malleable metals.

melting point (n) the temperature at which a solid changes into a liquid. Metals used in aircraft engines need to have very high melting points.

missile (n) a pilotless flying weapon which explodes when it hits its target.

monocoque construction (adj + n) a body, such as an egg, which has all the strength in the skin, without any internal reinforcement. Most aircraft have a semi-monocoque design so that the stresses are carried by the internal framework as well as the outer skin.

MRO (maintenance repair and overhaul) (n) The systematic and regular inspection, repair and parts replacement procedures to ensure that aircraft are airworthy.

multi-stage compressor (n) a compressor which consists of two or three compressors which act on the gas one after the other and increase the pressure more each time.

N

NDI (non-destructive inspection) (adj + n) careful checking of a material or component without taking it to pieces or removing a sample from it. Examples are eddy current examination of metal skins and the use of a special instrument called a boroscope to look inside engines.

nickel-cadmium battery (n) a type of battery made from cadmium and nickel hydroxide. It can be recharged and gives a steady supply of power without any reduction until just before it runs out.

nickel-metal-hydride battery (n) an improvement on the *nickel-cadmium battery*. One of these can store about 40% more power than an NC battery of the same size and weight.

nose-wheel (n) a landing and take-off wheel (often retractable) which is attached to the front part of the aircraft (the nose).

nozzle (n) the projecting end of a tube with a small hole in it which is used to spray liquids or gases.

O

O ring (n) a flexible circular washer, made of rubber, plastic or silicone, which has a circular cross-section. Often used for sealing high-pressure liquid or gas connections.

OBOGS (onboard oxygen generation system) (n) equipment which enables an aircraft to make its own oxygen as it flies, instead of loading and carrying large, heavy, pressurised containers.

oscillator (n) an electronic device which produces a high-frequency AC signal, either as current in a circuit or as radio waves.

oscilloscope (n) an electronic instrument which displays electronic signals visually on a screen. It is used for inspecting, testing and troubleshooting electronic circuits.

P

payload (n) the extra weight that an aircraft is carrying in the form of passengers, luggage, cargo or bombs.

piston (n) a solid cylinder that fits inside another cylinder and moves under pressure to push or displace gas or liquid.

pitch (n) 1. the angle at which the nose of an aircraft is pointing up or down; 2. the angle of the thread on a bolt or screw.

plated (adj) having a covering of one metal on the surface of another. Some electronic terminals are gold-plated to improve their electrical conductivity and resist corrosion.

pneumatic piston assembly (n) the cylinder and piston of an air pump.

potential freight load (n) the maximum weight of cargo that it is possible for an aircraft to carry, rather than what it actually carries normally.

power plant (**n**) the main engine or engines of an aircraft.

power to weight ratio (**n**) the amount of power an engine can produce compared with its weight. Jet engines have a higher PWR than internal combustion engines.

PPE (personal protective equipment) (**n**) clothes or equipment worn to prevent injuries or illness due to workplace hazards. Safety helmets, gloves and ear defenders are examples of PPE.

pressure transducer (**n**) a device which produces small electrical currents when it is put under mechanical pressure.

processor (**n**) a small electronic component which does calculations and sends and receives information inside a computer.

production cell (**n**) a small group of workers who are responsible for several stages in the production of a manufactured item such as a car or a plane. Some companies find that production cells work better than each worker having just one small repeated job.

propeller (**n**) a rotating device with two or more pitched blades which forces an aircraft through the air.

propeller shaft (**n**) The rotating shaft that the propeller is mounted on.

pushrod (**n**) a metal rod which connects a cam on the camshaft to a spring-loaded valve. The rotating cam pushes the rod against the valve to open it.

R

RADAR (radio detection and rangefinding) (**n**) equipment which sends very high-frequency radio signals to display the location and direction of a flying aircraft on a radar screen.

radiator (**n**) a form of heat exchanger designed for cooling liquids. A hot liquid circulates through the inside, and the large external surface area transmits unwanted heat into the air.

ram (**v & n**) (**v**) to suddenly increase pressure on a liquid solid or gas; (**n**) a device designed to do this. Air which is collected from the front

of a plane is called ram air.

reciprocal lathe (**n**) an early type of lathe which used backwards and forwards linear motion in the form of a bow, pole or strap to produce rotation of the workpiece.

reciprocating piston (**n**) a piston which moves constantly up and down in the cylinder of an engine, pump or compressor.

reduction gearbox (**n**) a set of mechanical gears designed to change the high speed of an engine into a lower speed that is more suitable for attached equipment. The propeller of an aircraft is often connected to a reduction gearbox.

reservoir (**n**) a container used to store a liquid or gas so that is always available when needed.

resistor (**n**) an electronic component with a value measured in ohms, which resists the flow of an electric current.

retract chamber (**n**) the lower part of a hydraulic cylinder. When it is filled with hydraulic fluid, the piston is forced to the top of the cylinder in order to retract an attached device.

rheostat (**n**) a variable resistor which is used to reduce the flow of electricity to equipment or circuits. Lights can be made dimmer or brighter by using a rheostat.

rib (**n**) a lateral beam or strut forming part of the framework of an aircraft.

rivet (**n**) a pin with two heads which is used to fasten metal parts very tightly together.

rocker arm (also called a rocker) (**n**) a lever in an internal combustion engine that connects a push rod to a valve.

rocket firing unit (**n**) part of the ejection seat mechanism that forces the seat and pilot out of the aircraft at extremely high speed.

rotary engine (**n**) an unusual type of internal combustion engine which has major parts that rotate in addition to or as well as the shaft. In the Wankel engine, the pistons rotate, and in the gnome aircraft engine, the engine block rotates around the shaft.

Take off **Glossary**

rotor (**n**) any device which turns round a centre point or axis.

runway (**n**) a long piece of ground with a prepared smooth surface which is used by aircraft for taking off and landing.

selector (**n**) a control switch, handle or knob which enables a pilot to choose a particular operation.

selector switch (**n**) an electrical selector.

semi-monocoque (**adj**) an aircraft design in which both the skin and the internal frame of the aircraft carry the stress loads. Also see *monocoque*.

sensor (**n**) an instrument which detects or reacts to a physical condition such as a rise in temperature or the presence of a dangerous gas. Aircraft have many different types of sensor to constantly monitor the engines, the fuel system and the cabin atmosphere.

simple transformer (**n**) a device consisting of a primary and secondary coil which changes the voltage of an AC current.

single-skin construction (**adj + n**) made with only one layer of covering material. Most modern aircraft are made with two skins with an insulating layer between them.

slip rings (**n**) rotating metal rings at the end of the shaft of an alternator which feed current to the supply circuit. They are similar to the commutator in a DC generator or motor.

solenoid (**n**) an *actuator* consisting of a moveable, soft, iron rod surrounded by a coil. When a current flows through the coil, the rod moves in or out to operate another device.

spark plug (**n**) a small device screwed into the top of the cylinder of an internal combustion engine. It produces a high-voltage spark which ignites the fuel mixture in the cylinder.

specific heat capacity (**n**) the amount of heat energy required to increase the temperature of a material one degree. The SHC of aluminium is about 900 jules per kg, whereas iron is about 470 jpk, which means that it takes twice as much energy to increase the temperature of aluminium as it does to increase the temperature of iron.

static charge (**n**) DC electricity which does not flow as a current but is released instantly. A capacitor can build up and store a static charge. Lightning is caused by a static charge of millions of volts.

stiffening (**n**) additional pieces of material fastened to a structural part such as a beam or strut to make it more rigid and less likely to bend under a load.

strainer (**n**) a type of filter designed to separate two or more materials from each other.

stressed skin panel (**n**) a piece of aircraft skin which is not just a cover, but is designed to withstand stress loads and contribute to the general strength of the structure.

stringer (**n**) a thin metal strip which goes along the length of the fuselage.

strut (**n**) a rod or beam which connects one part of a plane with another to increase strength and support. Small light aircraft often have struts between the wings and the fuselage.

sub-assembly (**n**) a group of connected components which are put together to make a part before it is fitted into a larger group *(assembly)* of connected parts.

sump (**n**) a *reservoir* of lubricating oil located at the bottom of an engine.

supercharger (**n**) a special high-speed fan which forces air into an internal combustion engine and increases its power.

taildragger (**n**) an informal name for an aircraft which has a landing wheel on the tail instead of the nose.

take-off selector (**n**) a *selector* used to set the control surfaces to the position required for the plane to leave the ground.

tensile strength (**n**) the ability of a material to withstand being pulled or stretched without breaking.

terminal block (**n**) a piece of insulating material containing metal connectors that the ends of wires are attached to.

thermal conductivity the property of allowing heat to pass through a material. Copper has a high thermal conductivity, whereas wood and polystyrene do not.

thrust (n) the pushing power of an aircraft engine which makes the plane move forward.

torsion (n) a twisting force which tries to turn something clockwise and anticlockwise at the same time.

tricycle (n) type of landing gear system found on most modern aircraft which consists of a nose wheel and two main wheels or sets of wheels in the middle of the aircraft, behind the centre of gravity.

trim (n) the balance of an aircraft in flight. To keep the aircraft in exactly the right position, small flight control surfaces called trim tabs have to be constantly adjusted.

turbo-fan engine (n) a type of jet engine, used on most large planes today, which is fitted with a large fan that provides most of the thrust.

turboprop engine (n) a type of jet engine which drives an external propeller.

turn coordinator (n) a flight instrument which tells the pilot how effectively and safely he is turning the aircraft in order to change direction.

V

vacuum (n) space or container where there is an absence of pressure and matter.

valve (n) a device that regulates the flow of gases, liquids or materials through a pipe or other passage.

volatility (n) the ability to change from a liquid to a vapour.

voltage probe (n) a thin conducting rod with a pointed end which can be put in contact with parts of a circuit that are difficult to reach during inspection and fault tracing.

voltage regulator (n) a device in a power supply system which ensures that current is supplied at a constant electromotive force (voltage), without sudden increases (surges) or decreases (drops).

voltmeter (n) an instrument which measures the strength of an electromotive force in volts.

W

workpiece (n) a piece of material that is being shaped, drilled, cut, filed or worked on in some way.

Electrical symbols

AC voltage source

Aerial/antenna

Autotransformer

Battery

Capacitor 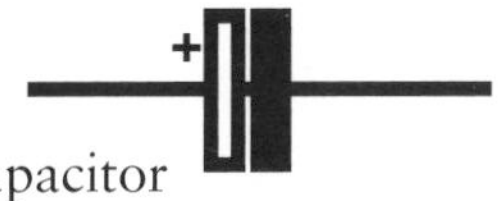   Polarised capacitor    Variable capacitor

Chassis earth

Circuit breaker

Coaxial cable

Diode

Double pole single throw switch

Double shielded conductor

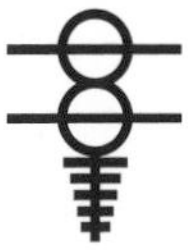

Earth

Fuse

Generator

Inductor  ⌒⌒⌒ Variable inductor

Iron core transformer

Lamp

Loudspeaker

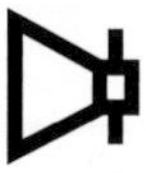

Motor

Piezoelectric crystal (transducer)

Pressure switch (closes on pressure release)

Relay

Relay switch

Resistor

Rheostat

Rotary switch

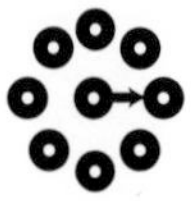

Shielded and twisted conductor

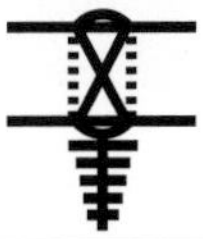

Shielded conductor

Single pole double throw normally closed, momentarily open switch

Single pole double throw switch

Single pole single throw switch

Solenoid

Spare conductor with insulated end

Switch

Terminal block

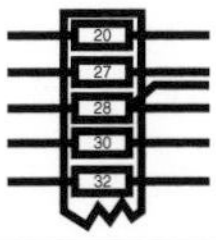

Terminal strip

Transformer

Wires cross (no contact) 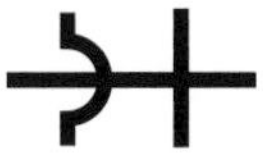   Wires cross (tie point)

Word list
Unit by unit

Unit 1

Lesson 1
air cushion vehicle (n)
bagless vacuum cleaner (n)
capacity (n)
gigabyte (n)
height (n)
horsepower (n)
innovative (adj)
kilometre (n)
LCD television (n)
memory (n)
MP3 player (n)
power (n)
research and development (n)
rotary engine (n)
speed (n)
technology (n)
vertical take-off (adj, n)
watt (n)
weight (n)

Lesson 2
aircraft (n)
component (n)
conditions (n)
crankshaft (n)
design (n)
drawing (n)
economics (n)
flywheel (n)
mass-production (n)
materials (n)
military (adj)
model (n)
prototype (n)
regulation (n)
sketch (n)
specification (n)

Lesson 3
alloy (n)
availability (n)
brittleness (n)
chemical stability (n)
coefficient of linear
    expansion (n)
composite (n)
corrode (v)
density (n)
ductility (n)
elasticity (n)
electrical resistivity (n)
hardness (n)
malleability (n)
mechanical (n)
melting point (n)
moulded plastic (n)
non-metallic (adj)
physical (adj)
plated metal alloy (n)
property (n)
range (n)
specific heat capacity (n)
specific operating
conditions (n)
strength (n)
suitability (n)
synthetic (adj)
thermal conductivity (n)
toughness (n)
transparent (adj)

Lesson 4
adhesive (adj, n)
airliner (n)
application (n)
characteristic (n)
commercial (adj)
fine (adj)

Pascal (n)
radial (adj)
tensile strength (n)

Lesson 5
enable (v)
facilities (n)
geometry (n)
horizontal (adj)
impression (n)
manufacturing costs (n)
optics (n)
orthographic projection (n)
passenger (n)
perspective (n)
pictorial (adj)
point of view (n)
potential freight load (n)
safety (n)
sponsor (n, v)
technique (n)
three-dimensional (adj)
two-/three-point
perspective (n)
two-dimensional (adj)
vanishing point (n)
visualise (v)

Lesson 6
acceleration (n)
cabin (n)
ceiling (n)
circumference (n)
crew (n)
cubic metre (n)
distance (n)
divided (adj)
duration (n)
flight (n)
kilogram (n)

kilometres per hour (n)
length (n)
maximum (adj, n)
minus
multiplied
payload (n)
plus power plant (n)
relative to sphere (n)
surface area (n)
take-off (n)
volume (n)
width (n)

Lesson 7
air brake (n)
decoy (n)
display (adj, n)
ejection seat (n)
foreplane (n)
fuel (n)
fuel tank (n)
fuselage (n)
long-range (adj)
medium-range (adj)
missile (n)
port (adj)
radar (n)
refueling probe (n)
retractable (adj)
short-range (adj)
starboard (adj)
tail (n)
tank (n)
turbofan engine (n)
undercarriage (n)
wing (n)

Lesson 8
brake (n)
economics (n)

funding (n)
hangar (n)
invest (v)
manufacturing (n)
runway (n)
scrapped (adj)
trial (n, v)

Lesson 9

approximately (adv)
area (n)
centimetre (n)
cruise speed (n)
customer (n)
data (n)
dimension (n)
distance (n)
equivalent (adj)
estimate (v)
feet (n)
imperial (adj)
inch (n)
kilowatt (n)
metre (n)
metric (adj)
miles per hour (n)
performance (n)
pound (n)
rough (adj)
square (adj, n)
war plane (n)
wingspan (n)

Unit 2

Lesson 1

bend (v)
blade (n)
calibrate (v)
edge (n)
face (n)
file (n, v)
grip (v)
hammer (n, v)

hand tool (n)
handle (n)
jaw (n)
measure (v)
pliers (n)
pound (v)
power tool (n)
precision (n)
punch (n, v)
saw (n, v)
screw (n, v)
screwdriver (n)
sharp (adj)
sharpen (v)
size (n)
teeth (n)
thread (n)
tighten (v)
vice (n)

Lesson 2

batch (n)
beat (v)
blade (n)
flexible (adj)
fold (n, v)
forge (v)
gap (n)
machine tool (n)
molten (adj)
polish (v)
process (n)
razor-sharp (adj)
sheet (n)
smith (n)
solid (adj)
tie (v)
weld (v)
welding (n)
wrench (n)
wrought iron (n)

Lesson 3

accuracy (n)
axis/axes (n)

CNC machine tool (n)
configuration (n)
cutter (n)
hand-operated (adj)
interchangeable (adj)
key (n)
margin of error (n)
milling machine (n)
operation (n)
repeatability (n)
tailor (v)
tolerance (n)
versatile (adj)

Lesson 4

adjust (v)
air-powered (adj)
align (v)
base (n)
drill bit (n)
feed (n)
gearbox (n)
handwheel (n)
pillar (n)
pneumatic (adj)
transmit (v)
turning tool (n)
work table (n)
workpiece (n)

Lesson 5

bow (n)
crank (n)
gyroscope (n)
internal combustion engine (n)
irrigation (n)
lathe (n)
momentum (n)
pole (n)
reciprocal lathe (n)
rotation (n)
strap (n)
treadle (n)
wood turning (n)

Lesson 6

apron (n)
connect (v)
dial (n)
ensure (v)
feed (n, v)
headstock (n)
isolator (n)
knob (n)
lead screw (n)
lever (n)
longitudinal axis (n)
mains (n)
protection (n)
reverse (adj, n, v)
selector (n)
spindle (n)
thread pitch (n)
transverse axis (n)

Lesson 7

aluminium (n)
asbestos (n)
chromium (n)
coating (n)
conductivity (n)
conductor (n)
construction (n)
copper (n)
corrode (v)
corrosion (n)
durability (n)
electrochemically (n)
gold (n)
lead (n)
lightness (n)
lightweight (adj)
manganese (n)
oxide (n)
resistance (n)
stainless steel (n)
versatile (adj)
zinc (n)

Lesson 8

burr (n)

hole (n)

instructor (n)

malleable (adj)

mark out (v)

radius (n)

scribe (v)

shears (n)

smooth (adj)

Lesson 9

airworthy (adj)

bar (n)

blind rivet (n)

bolt (n)

clip (n)

fastening (n)

insert (v)

man-hour (n)

non-structural part (n)

pin (n)

pinhead (n)

pre-drilled hole (n)

rivet (n)

riveting tool (n)

shear (adj)

shearing strength (n)

standard rivet (n)

structural part (n)

titanium (n)

vibration (n)

washer (n)

Lesson 10

factory (n)

kit (n)

tin snip (n)

Unit 3

Lesson 1

blade (n)

control surface (n)

fin (n)

flap (n)

hinged (adj)

horizontal (adj)

landing gear (n)

longitudinal (adj)

main body (n)

pitched (adj)

power plant (n)

propeller (n)

rod (n)

rudder (n)

skin (n, v)

structure (n)

strut (n)

support (n, v)

surface (n)

vertical (adj)

Lesson 2

backwards (adj, adv)

bank (v)

bending (n)

climb (v)

clockwise (adj, adv)

compression (n)

dive (v)

fall (v)

force (n)

forwards (adj, adv)

gravity (n)

hover (v)

lift (n, v)

roll (v)

rotate (v)

sideways (adj, adv)

stress (n)

tension (n)

torsion (n)

Lesson 3

descent (n)

gain height (v)

grab (v)

manoeuvre (n, v)

pull up (v)

shallow angle (n)

turn around (v)

twist (v)

wingspan (n)

Lesson 4

airframe (n)

function (n)

hollow (adj)

internal [structure] (n)

rib (n)

rigid (adj)

skeleton (n)

withstand (v)

Lesson 5

bulkhead (n)

extend (v)

external (adj)

frame (n)

framework (n)

lateral (adj)

lengthwise (adj)

longeron (n)

monocoque (adj)

plate (n)

semi-monocoque (adj)

spar (n)

stiffening (n)

stressed skin panel (n)

stringer (n)

trussed (adj)

Lesson 6

assembly (n)

attach (v)

carry out (v)

electronics (n)

engine (n)

fit (v)

flight control system (n)

flight test (n)

hydraulics (n)

install (v)

main assembly (n)

perform (v)

pre-flight test (n)

production line (n)

splice (v)

weapons system (n)

Lesson 7

bridge crane (n)

hangar (n)

in situ (adv)

part (n)

production site (n)

shift (n)

site (n)

sub-assembly (n)

tarmac (n)

Lesson 8

aerospace industry (n)

automation (n)

contract (n)

inaccessible (adj)

maintenance (n)

manual procedure (n)

manufacturing industries (n)

robot technology (n)

robotic assembly (n)

security (n)

single-skin construction (n)

wing box (n)

Lesson 9

background (n)

draughtsman (n)

engineer (n)

experience (n)

machinist (n)

manual (n)

modification (n)

navigation (n)

qualification (n)

steering (n)

systems engineer (n)

technician (n)

Lesson 10

be accountable for
    [something] (v)

brief [someone] (v)

cellular manufacturing (n)

chain of communication (n)

clarification (n)

collective responsibility (n)

communication (n)

intercom (n)

log (n)

malfunction (n, v)

production cell (n)

relieve [someone] (v)

terminology (n)

Unit 4

Lesson 1

channel (n)

constant (adj)

flap valve (n)

flow rate (n)

gradient (n)

inlet (adj, n)

kinetic energy (n)

level (n)

outlet (adj, n)

pipe (n)

piston (n)

pool (n)

potential energy (n)

pressure (n)

pump (n, v)

reciprocating piston (n)

reservoir (n)

slope (n)

sophisticated (adj)

source (n)

suction (n)

supply (n, v)

Lesson 2

clack valve (n)

cycle (n)

delivery valve (n)

device (n)

fluid (n)

hydraulic (adj)

increase (n, v)

primary flow (n)

ram (n, v)

spring (n)

spring-loaded (adj)

upwards (adv)

valve (n)

Lesson 3

atmospheric (adj)

boiling point (n)

freezing point (n)

hydraulic power (n)

hydraulic system (n)

lifting (n)

linkage (n)

loading (n)

mechanical advantage (n)

tarmac (n)

technological (adj)

towing (n)

Lesson 4

actuate (v)

actuator (n)

control (v)

de-energise (v)

energise (v)

extend (v)

indicate (v)

microswitch (n)

operate (v)

psi [pounds per square
    inch] (n)

relay (n, v)

retract (v)

select (v)

selector switch (n)

solenoid (n)

Lesson 5

absorb [shock] (v)

artificial horizon (n)

descent (n)

detect (v)

determine (v)

distribute [the load] (v)

emergency landing (n)

flexibility (n)

impact (n)

line up (v)

load (n)

minimise (v)

mobility (n)

orientation (n)

precise (adj)

pressurised fluid (n)

shock (n)

simultaneous (adj)

spread (v)

Lesson 6

air resistance (n)

center of gravity (n)

groundloop (n, v)

isolator (n)

leg (n)

nose wheel (n)

rough ground (n)

spin round (v)

stationary (adj)

taildragger (n)

tricycle (n)

visibility (n)

Lesson 7

chamber (n)

control angle (n)

drag (n)

inoperative (adj)

intermediate chamber (n)

land selector (n)

partially/fully extended (adj)

piston head chamber (n)

primary (adj)

redirect (v)

retract chamber (n)

secondary (adj)

spring-loaded (adj)

take-off selector (n)

trainer (n)

Lesson 8

aileron (n)

cable (n)

elevator (n)

joystick (n)

lateral axis (n)

longitudinal axis (n)

pedal (n)

pitch (n, v)

pre-flight check (n)

roll (n, v)

slow down (v)

vertical axis (n)

yaw (n, v)

Lesson 9

buckle (v)

control column (n)

excessive load (n)

groove (n)

hollow (adj)

play (n)

pulley [wheel] (n)

rigid (adj)

rotor (n)

slack (adj)

stiff (adj)

tension (n)

terminal (n)

torque (n)

tube (n)

Unit 5

Lesson 1
alternator (n)
camshaft (n)
cylinder block (n)
engine block (n)
fan (n)
fan belt (n)
filter (n)
generator (n)
pushrod (n)
rocker (n)
rocker arm (n)
spark plug (n)
starter motor (n)
strainer (n)
sump (n)
timing chain/belt (n)

Lesson 2
cruise (v)
gas turbine (n)
grounded (adj)
in-line (adj)
price (n)
reliability (n)
rubber band (n)
simplicity (n)
spare part (n)
streamlined (adj)

Lesson 3
angular (n)
decrease (v)
increase (v)
linear (adj)
mass (n)
overall (adj, adv)
perpendicular (adj)
skater (n)
stable (adj)
straight line (n)
strain (n, v)

top (n)
velocity (n)

Lesson 4
admit (v)
air-cooled (adj)
by hand (adv)
downwards (adv)
exhaust (n)
expand (v)
ignited gas (n)
mixture (n)
power stroke (n)
power-to-weight ratio (n)
radiator (n)
ratio (n)
resist (v)
suck [in] (v)
vacuum (n)

Lesson 5
automatic (adj)
aviation fuel (n)
bearings (n)
bore (n)
breakdown (n)
data (n)
diameter (n)
displacement (n)
gear (n)
high gear (n)
liner (n)
low gear (n)
lubrication (n)
power rating (n)
starter (n)
stroke (n)
supercharger (n)

Lesson 6
assemble (v)
blocked (adj)
brass (n)
clean (v)

cylinder head gasket (n)
damaged (adj)
dirty (adj)
dismantle (v)
flood (n, v)
frame (n)
gasket (n)
inspect (v)
leak (n, v)
loose (adj)
machine (v)
mount (v)
out of alignment (adj)
overhaul (v)
plug (n)
reattach (v)
rebuild (v)
refit (v)
remove (v)
replace (v)
sandblast (v)
strip down (v)
transfer (v)

Lesson 7
air intake (n)
act on (v)
add [to] (v)
combustion chamber (n)
discharge (v)
draw in (v)
drive (v)
exert pressure on (v)
exhaust duct (n)
expel (v)
mix (v)
multistage compressor (n)
propeller shaft (n)
provide [power] (v)
reduction gearbox (n)
spark (n)
turboprop (n)

Lesson 8
altitude (n)
bypass (v)
core (n)
cowling (n)
cruising altitude (n)
exit velocity (n)
extract (v)
fahrenheit (n)
jet fighter (n)
revolutions per minute (n)
statistics (n)
thrust (n)

Lesson 9
chemical reaction (n)
corrosiveness (n)
efficiency (n)
flash point (n)
freezing point (n)
kerosene (n)
liquid (n)
lubricity (n)
net [heat] (adj)
standard specification (n)
sulphur (n)
volatility (n)

Lesson 10
battery (n)
booster pump (n)
centrifuge (n)
collector tank (n)
delivery (n)
flammable (adj)
flange (n)
freeze (v)
gear pump (n)
jet pump (n)
seal (n, v)
transfer (n)
vane (n)
venturi principle (n)

Unit 7

Lesson 1

assess (v)

caution (n, v)

consult (v)

evaluate (v)

first aid (n)

forbid (v)

hazard (n)

maintain (v)

mandatory (adj)

manual (adj)

notice (n)

PPE (n)

prevent (v)

prohibit (v)

risk (n, v)

safety equipment (n)

scald (n, v)

sign (n)

warning (adj, n)

Lesson 2

change (= replace) (v)

clean up (v)

disconnect (v)

dry out (v)

fumes (n)

guard (n)

ignition (n)

switch on/off (v)

well-ventilated (adj)

Lesson 3

blanket (n)

colour-coded (adj)

conduct (v)

emergency (n)

extinguish (v)

film (n)

fire extinguisher (n)

foam (n)

fuel (v)

multi-purpose (adj)

powder (n)

residue (n)

suitable (adj)

versatile (adj)

Lesson 4

breathing equipment (n)

bumpy (adj)

check (n, v)

Civil Aviation Authority (n)

cockpit (n)

confidential (adj)

evacuation (n)

fray (v)

harness (n)

intercom (n)

life jacket (n)

megaphone (n)

oxygen mask (n)

passenger cabin (n)

priority (n)

safety inspector (n)

seat belt (n)

service tag (n)

sick bag (n)

turbulence (n)

twisted (adj)

Lesson 5

bracket (n)

chafing (n)

contamination (n)

cracking (n)

discard (v)

distortion (n)

elongation (n)

evidence (n)

fading (n)

finish (n)

fixing (n)

loosen (n)

mechanism (n)

repair (n, v)

report (v)

restraint (n)

saddle washer (n)

scrutinize (v)

stitch/stitching (n)

strap (n)

tear (n, v)

wear (v)

Lesson 6

agent (n)

atmosphere (n)

burning [process] (adj)

damage (n)

disperse (v)

displace (v)

environment (n)

environmentally friendly (adj)

fraction (n)

halon (n)

harm (n, v)

interrupt (v)

ozone layer (n)

put out [a fire] (v)

react [with chemicals] (v)

spray (n, v)

suffocate (v)

Lesson 7

ballistic threat (n)

built-in (adj)

eliminate (v)

enriched (adj)

ground support (n)

HEI (high explosive
    incendiary) (adj)

incendiary round (n)

inert (adj)

integral inlet (adj)

life cycle (adj, n)

lightning (n)

nitrogen (n)

slosh (v)

static discharge (n)

survivability (n)

threat (n)

ullage (n)

vapour (n)

vulnerability (n)

Lesson 8

back pressure (n)

block (v)

braided (adj)

clamp (n, v)

concentration (n)

concentrator (n)

crimped (adj)

defect (n)

deteriorate (v)

downstream (adv)

hose (n)

hypoxia (n)

lethal (adj)

line [= tube] (n)

LOX converter (n)

message traffic (n)

OBOGS (onboard oxygen
    generating system) (n)

quality assurance (n)

sleeve (n)

Lesson 9

activate (v)

breech (adj, v)

collide (v)

deploy (v)

drogue parachute (n)

ejection gun (n)

initiate (v)

limb-restraint cord (n)

multi-tubed (adj)

navigator (n)

parachute (n)

propel (v)

reinforce (v)

remote (adj)

rocket firing unit (n)

rocket motor (n)

stabilise (v)

static line (n)

telescopic (adj)

Lesson 10

accumulator (n)

backup system (n)

charge (v)

check valve (n)

emergency package (n)

oil head (n)

pressure transducer (n)

release valve (n)

restrictor (n)

survival (n)

take over (n, v)

Unit 8

Lesson 1

bead (n)

bulge (n, v)

bump (v)

burst (v)

crosswind (n)

deflate (v)

flip (v)

foreign body (n)

gauge (n)

groove (n)

inflated (adj)

layer (n)

misalignment (n)

sidewall (n)

skid (v)

tread (n)

tyre (n)

Lesson 2

absorb (v)

annular space (n)

axle (n)

compressed (adj)

cylinder barrel (n)

dampen (n)

displaced (adj)

landing gear leg (n)

non-level (adj)

phase (n)

pneumatic (adj)

reciprocation (n)

recoil (v)

shock (n)

slip ring (n)

uneven (adj)

Lesson 3

basket (n)

burner (n)

Celsius/centigrade (n)

copilot (n)

dense (adj)

envelope (n)

equivalent (adj, adv)

Kelvin (n)

occupy (v)

propane (n)

proportional to (adv)

Lesson 4

arm (v)

bypass (adj, n)

cock (v)

compressor (n)

housing (n)

maintenance manual (n)

oil cooler (n)

pneumatic piston
    assembly (n)

prime (v)

purge (v)

set (v)

shutdown (adj)

tattle tale (n)

vent (adj, n)

Lesson 5

circulate (v)

compression [stroke] (adj, n)

condense (v)

condenser (n)

evaporate (v)

evaporator (n)

freeze (v)

give off (v)

induction [stroke] (adj, n)

liquefy (v)

melt (v)

orifice (n)

piping (n)

refrigerant (n)

solidify (v)

state (v)

stroke (n)

vaporise (v)

waste (v)

Lesson 6

absolute humidity (n)

absorption (n)

atmospheric humidity (n)

bulb (n)

condensation (n)

controlling humidity (n)

humidify (v)

humidity (n)

hygrometer (n)

intermix (v)

measuring humidity (n)

moist (adj)

moisture (n)

molecule (n)

relative humidity (n)

saturated (adj)

saturation point (n)

temperature reading (n)

Lesson 7

AC (air conditioning) (n)

activate (v)

coil (n)

cooler (n)

draw in (v)

filter (n, v)

freshen (v)

hygrostat (n)

recirculate (v)

reheater (n)

thermostat (n)

unit (n)

Lesson 8

consume (v)

contaminant (n)

dilute (v)

equipped (adj)

high-efficiency (adj)

interval (n)

jetliner (n)

replenish (v)

ventilate (v)

Lesson 9

distribute (v)

draw off (v)

duct (n)

exhausted (adj)

grille (n)

heat exchanger (n)

intake (n)

lobe (n)

microscopic particle (n)

outflow valve (n)

overhead [outlet] (adj)

pattern (n)

trap (v)

Lesson 10

bleed (adj, v)

defogging nozzle (n)

ECS [environmental
    control system] (n)

extract (v)

firewall (n)

nozzle (n)

outlet (n)

sensor (n)

separator (n)

turbine (n)

Unit 9

Lesson 1

ambient (adj)
armour (n)
bedding (n)
busbar (n)
cross-sectional area (n)
current (n)
dissipate (v)
filler (n)
heavy-duty (adj)
insulate (v)
insulation sleeve (n)
multi-cored (adj)
sector-shaped (adj)
sheath (n)
tape (n)
voltage (n)
wire (n)

Lesson 2

ampere (n)
bundled [cable] (adj)
circuit (n)
conduit (n)
curve (n)
drop (n, v)
formula (n)
ohm (n)
square root (n)
value (n)
volt (n)
voltage (n)
watt (n)

Lesson 3

cable clamp (n)
drainage hole (n)
grommet (n)
installation (n)
loop (n)
lug (n)
maintenance check (n)
maintenance form (n)

protrude (v)
redo (v)
reposition (v)
ring (n)
slack (adj, n)
terminal block (n)

Lesson 4

lead-acid [battery] (adj, n)
lithium-ion [battery] (adj, n)
nickel-cadmium
    [battery] (adj, n)
nickel-metal-hydride (adj, n)
percentage capacity (n)
recharge (v)
self-discharge rate (n)
storage (n)

Lesson 5

charge up [battery] (v)
cut out [engine] (v)
EPU [external power unit] (n)
generator (n)
handpropping (n)
injury (n)
risky (adj)
run down [battery] (v)
taxi off (v)
turn over [engine] (v)

Lesson 6

[radio] static (n)
air gap (n)
armature (n)
brush (n)
brush holder (n)
build up (v)
burn out (v)
capacitance meter (n)
capacitor (n)
commutator (n)
contact (v)
electrical arcing
    (sparking) (n)
multi-graph (n)

operating manual (n)
pitted (adj)
scale (n)
shorted (adj)
snag (v)
tangled (adj)

Lesson 7

circuit diagram (n)
diode (n)
electrolytic capacitor (n)
filament bulb (n)
inductor (n)
lamp (n)
resistor (n)
semiconductor (n)
simple transformer (n)
single pole switch (n)
troubleshooting (n)
voltmeter (n)

Lesson 8

AC (alternating current) (n)
analogue multimeter (n)
cable pulling draw tape (n)
crimping tool (n)
DC (direct current) (n)
digital multimeter (n)
electrical tool catalogue (n)
hacksaw (n)
mains (adj, n)
multi-grip (adj)
side cutters (n)
socket set (n)
solder (n, v)
soldering gun (n)
soldering iron (n)
tape measure (n)
voltage probe (n)
WD40 lubricant (n)
wire strippers (n)

Lesson 9

avionics (n)
bring online (v)

display panel (n)
dual output (n)
dual-role (adj)
hertz (n)
inverter (n)
isolate (v)
power socket (n)
starter-generator (n)
transistorised static
    inverter (n)
trickle-charge (v)
voltage regulator (n)

Lesson 10

circuit-breaker (n)
common connection
    point (n)
electromagnet (n)
flat strip (n)
fuse (n)
hollow (adj)
melt (v)
multiple switch (n)
on-board (adj)
overloaded (adj)
relay (n)
remote switch (n)

Unit 10

Lesson 1

air traffic controller
    [ATC] (n)
attitude (n)
collision (n)
departure point (n)
flight data recorder (n)
trim (n)

Lesson 2

airborne (adj)
altimeter (n)
back-up (v)
barometer (n)

button (n)
compass (n)
gyro-compass (n)
panel (n)
recalibrate (v)
stall (v)
turn-coordinator (n)

Lesson 3
ammeter (n)
analogue (adj)
digital (adj)
division (n)
input (n)
instrumentation (n)
needle (n)
output (n)
pointer (n)
solid state (adj)

Lesson 4
algebra (n)
application (n)
arithmetic (n)
calculus (n)
chips (n)
circuit board (n)
coordinate (v)
electromechanical (adj)
fabricate (v)
fine-tune (v)
functional (adj)
hardware (n)
integral (adj)
interpret [data] (v)
junction box (n)
know-how (n)
layout drawing (n)
magneto (n)
malfunction (n)
optimum (adj)
oscilloscope (n)
personnel (adj)
processor (n)
programming (n)

regulate (v)
schematic (n)
set up (v)
software (n)
systemic [problems] (adj)
troubleshoot (v)

Lesson 5
amplifier (n)
chassis (n)
click (n)
gyro-system (n)
hum (n)
interference (n)
rheostat (n)

Lesson 6
fault (n)
integrated circuit (n)
intermittent (adj)
magnifying glass (n)
mishandle (v)
parallel (adj)
PCB (printed circuit
    board) (n)
reading (n)
right angles (adj)
trace [fault] (v)

Lesson 7
aerosol solvent cleaner (n)
foreign matter (n)
grease (n)
heat sink (n)
lacquer (n)
particle (n)
short circuit (n, v)
solder sucker (n)
trim (v)

Lesson 8
aerial (n)
air to ground
    communication (n)
antennas (n)

clear for take-off (v)
dots and dashes (n)
echo (n)
flight path (n)
frequency (n)
identification code (n)
inbound (adj)
landmark (n)
Morse code (n)
navigation chart (n)
non-precision aid (n)
outbound (adj)
path (n)
quadrant (n)
RADAR [Radio Detection
    and Ranging] (n)
radio-based system (n)
refine (v)
tower [radio] (n)
transmitter receiver (n)
two-way radio (n)
VOR [Very high-frequency
    Omni-directional Radio
    range] (n)

Lesson 9
ELT [emergency locator
    transmitter] (n)
format (v)
GPS [Global Positioning
    System] (n)
grid (n)
icon (n)
latitude (n)
longitude (n)
monitor (n)
radar tracking (n)
satellite (n)
update (n)

Lesson 10
AC-excited (adj)
crack (n)
eddy current (n)
grain size [of metals] (n)

induce (v)
magnetic field (n)
oscillator (n)
permeability (n)
probe (n)
void (n)

Unit 11

Lesson 1
airworthiness (n)
appropriate (adj)
authority (n)
certify (v)
factor (n)
regulatory (adj)
shaft (n)

Lesson 2
compulsory (adj)
deactivate (v)
flatten (v)
live (adj)
personnel (n)
reset (v)
secure (adj)
tags [warning ~] (n)

Lesson 3
complex (adj)
disassemble (v)
nut (n)
o-ring (n)
put back (in, together,
    etc.) (v)
reassemble (v)
reinstall (v)
take apart (v)

Lesson 4
apply (v)
contaminate (v)
improvement (n)

Lesson 5

coning (n)

disc (n)

field conditions (n)

industrial pollution (n)

lining (n)

NDI (non-destructive
    inspection) (n)

restore (v)

sand (v)

sandpaper (n)

schedule (n, v)

scratch (n, v)

thickness (n)

Lesson 6

access door (n)

indicator (n)

lock (n)

locking pin (n)

miss [the problem] (v)

spot [the problem] (v)

verify (v)

Lesson 7

access equipment (n)

adjacent (adj)

clearance (n)

extension lead (n)

jack (n)

platform (n)

remote control unit (n)

servicing equipment (n)

work cage (n)

Lesson 8

docking (n)

domestic water service (n)

fire detection (v)

process water service (n)

steady (v)

take out [of service] (v)

undertake [maintenance] (v)

works [= maintenance] (n)

wrap around (v)

Lesson 9

abrasive (adj)

castor (n)

container (n)

dispose of (v)

drain (n, v)

failure (n)

loss of load (n)

MRO (Maintenance
    Repair and Overhaul)
    saddle (n)

Lesson 10

authorised (adj)

discrepancy (n)

document (v)

IAW (in accordance with)

L/H (left-hand)

R/H (right-hand)

record (v)

sign off (= authorise) (v)

work order (n)

Word list
Alphabetical

abrasive (adj)

absolute humidity (n)

absorb (v)

absorption (n)

AC (air conditioning) (n)

AC (alternating current) (n)

acceleration (n)

access door (n)

access equipment (n)

accumulator (n)

accuracy (n)

AC-excited (adj)

act on (v)

activate (v)

actuate (v)

actuator (n)

add [to] (v)

adhesive (adj, n)

adjacent (adj)

adjust (v)

admit (v)

aerial (n)

aerosol solvent cleaner (n)

aerospace industry (n)

agent (n)

aileron (n)

air brake (n)

air cushion vehicle (n)

air gap (n)

air resistance (n)

air to ground communication (n)

air traffic controller [ATC] (n)

airborne (adj)

air-cooled (adj)

aircraft (n)

airframe (n)

airliner (n)

air-powered (adj)

airworthiness (n)

airworthy (adj)

algebra (n)

align (v)

alloy (n)

alternator (n)

altimeter (n)

altitude (n)

aluminium (n)

ambient (adj)

ammeter (n)

ampere (n)

amplifier (n)

analogue (adj)

analogue multimeter (n)

angular (n)

annular space (n)

antennas (n)

application (n)

apply (v)

appropriate (adj)

approximately (adv)

apron (n)

area (n)

arithmetic (n)

arm (v)

armature (n)

armour (n)

artificial horizon (n)

asbestos (n)

assemble (v)

assembly (n)

assess (v)

atmosphere (n)

atmospheric (adj)

atmospheric humidity (n)

attach (v)

attitude (n)

authorised (adj)

authority (n)

automatic (adj)

automation (n)

availability (n)

aviation fuel (n)

avionics (n)

axis/axes (n)

axle (n)

back pressure (n)

background (n)

back up (v)

backup system (n)

backwards (adj, adv)

bagless vacuum cleaner (n)

ballistic threat (n)

bank (v)

bar (n)

barometer (n)

base (n)

basket (n)

batch (n)

battery (n)

be accountable for [something] (v)

bead (n)

bearings (n)

beat (v)

bedding (n)

bend (v)

bending (n)

blade (n)

blanket (n)

bleed (adj, v)

blind rivet (n)

block (v)

blocked (adj)

boiling point (n)

bolt (n)

booster pump (n)

bore (n)

bow (n)

bracket (n)

braided (adj)

brake (n)

brass (n)

breakdown (n)

breathing equipment (n)

breech (adj, v)

bridge crane (n)

brief [someone] (v)

bring online (v)

brittleness (n)

brush (n)

brush holder (n)

buckle (v)

build up (v)

built-in (adj)

bulb (n)

bulge (n, v)

bulkhead (n)

bump (v)

bumpy (adj)

bundled [cable] (adj)

burn out (v)

burner (n)

burning [process] (adj)

burr (n)

burst (v)

busbar (n)

button (n)

by hand (adv)

bypass (adj, n, v)

cabin (n)

cable (n)

cable pulling draw tape (n)

calculus (n)

calibrate (v)

camshaft (n)

capacitance meter (n)

capacitor (n)

capacity (n)

carry out (v)

castor (n)

caution (n, v)

ceiling (n)

cellular manufacturing (n)

Celsius/centigrade (n)

center of gravity (n)

centimetre (n)

centrifuge (n)

certify (v)

chafing (n)

chain of communication (n)

chamber (n)

change (= replace) (v)

channel (n)

characteristic (n)

charge (v)

charge up [battery] (v)

chassis (n)

check (n, v)

check valve (n)

chemical reaction (n)

chemical stability (n)

chips (n)

chromium (n)

circuit (n)

circuit board (n)

circuit diagram (n)

circuit-breaker (n)

circulate (v)

circumference (n)

Civil Aviation Authority (n)

clack valve (n)

clamp (n, v)

clarification (n)

clean (v)

clean up (v)

clear for take-off (v)

clearance (n)

click (n)

climb (v)

clip (n)

clockwise (adj, adv)

CNC machine tool (n)

coating (n)

cock (v)

cockpit (n)

coefficient of linear expansion (n)

coil (n)

collective responsibility (n)

collector tank (n)

collide (v)

collision (n)

colour-coded (adj)

combustion chamber (n)

commercial (adj)

common connection point (n)

communication (n)

commutator (n)

compass (n)

complex (adj)

component (n)

composite (n)

compressed (adj)

compression (n)

compression [stroke] (adj, n)

compressor (n)

compulsory (adj)

concentration (n)

concentrator (n)

condensation (n)

condense (v)

condenser (n)

conditions (n)

conduct (v)

conductivity (n)

conductor (n)

conduit (n)

confidential (adj)

configuration (n)

coning (n)

connect (v)

constant (adj)

construction (n)

consult (v)

consume (v)

contact (v)

container (n)

contaminant (n)

contaminate (v)

contamination (n)

contract (n)

control (v)

control angle (n)

control column (n)

control surface (n)

controlling humidity (n)

cooler (n)

coordinate (v)

copilot (n)

copper (n)

core (n)

corrode (v)

corrosion (n)

corrosiveness (n)

cowling (n)

crack (n)

cracking (n)

crank (n)

crankshaft (n)

crew (n)

crimped (adj)

crimping tool (n)

cross-sectional area (n)

crosswind (n)

cruise (v)

cruise speed (n)

cruising altitude (n)

cubic metre (n)

current (n)

curve (n)

customer (n)

cut out [engine] (v)

cutter (n)

cycle (n)

cylinder barrel (n)

cylinder block (n)

cylinder head gasket (n)

D

damage (n)

damaged (adj)

dampen (n)

data (n)

DC (direct current) (n)

deactivate (v)

decoy (n)

decrease (v)

de-energise (v)

defect (n)

deflate(v)

defogging nozzle (n)

delivery (n)

delivery valve (n)

dense (adj)

density (n)

departure point (n)

deploy (v)

descent (n)

design (n)

detect (v)

deteriorate (v)

determine (v)

device (n)

dial (n)

diameter (n)

digital (adj)

digital multimeter (n)

dilute (v)

dimension (n)

diode (n)

dirty (adj)

disassemble (v)

disc (n)

discard (v)

discharge (v)

disconnect (v)

discrepancy (n)

dismantle (v)

disperse (v)

displace (v)

displaced (adj)

displacement (n)

display (adj, n)

display panel (n)

dispose of (v)

dissipate (v)

distance (n)

distortion (n)

distribute (v)

distribute [the load] (v)

dive (v)

divided (adj)

division (n)

docking (n)

document (v)

domestic water service (n)

dots and dashes (n)

downstream (adv)

downwards (adv)

drag (n)

drain (n, v)

drainage hole (n)

draughtsman (n)

draw in (v)

draw off (v)

drawing (n)

drill bit (n)

drive (v)

drogue parachute (n)

drop (n, v)

dry out (v)

dual output (n)

dual-role (adj)

duct (n)

ductility (n)

durability (n)

duration (n)

echo (n)

economics (n)

ECS [environmental control system] (n)

eddy current (n)

edge (n)

efficiency (n)

ejection gun (n)

ejection seat (n)

elasticity (n)

electrical arcing (sparking) (n)

electrical resistivity (n)

electrical tool catalogue (n)

electrochemically electrolytic capacitor (n)

electromagnet (n)

electromechanical (adj)

electronics (n)

elevator (n)

eliminate (v)

elongation (n)

ELT [emergency locator transmitter] (n)

emergency (n)

emergency landing (n)

emergency package (n)

enable (v)

energise (v)

engine (n)

engine block (n)

engineer (n)

enriched (adj)

ensure (v)

envelope (n)

environment (n)

environmentally friendly (adj)

EPU [external power unit] (n)

equipped (adj)

equivalent (adj, adv)

estimate (v)

evacuation (n)

evaluate (v)

evaporate (v)

evaporator (n)

evidence (n)

excessive load (n)

exert pressure on (v)

exhaust (n)

exhaust duct (n)

exhausted (adj)

exit velocity (n)

expand (v)

expel (v)

experience (n)

extend (v)

extension lead (n)

external (adj)

extinguish (v)

extract (v)

fabricate (v)

face (n)

facilities (n)

factor (n)

factory (n)

fading (n)

fahrenheit (n)

failure (n)

fall (v)

fan (n)

fan belt (n)

fastening (n)

fault (n)

feed (n, v)

feet (n)

field conditions (n)

filament bulb (n)

file (n, v)

filler (n)

film (n)

filter (n, v)

fin (n)

fine (adj)

fine-tune (v)

finish (n)

fire detection (v)

fire extinguisher (n)

firewall (n)

first aid (n)

fit (v)

fixing (n)

flammable (adj)

flange (n)

flap (n)

flap valve (n)

flash point (n)

flat strip (n)

flatten (v)

flexibility (n)

flexible (adj)

flight (n)

flight control system (n)

flight data recorder (n)

flight path (n)

flight test (n)

flip (v)

flood (n, v)

flow rate (n)

fluid (n)

flywheel (n)

foam (n)

fold (n, v)

forbid (v)

force (n)

foreign body (n)

foreign matter (n)

foreplane (n)

forge (v)

format (v)

formula (n)

forwards (adj, adv)

fraction (n)

frame (n)

framework (n)

fray (v)

freeze (v)

freezing point (n)

frequency (n)

freshen (v)

fuel (v)

fuel tank (n)

fumes (n)

function (n)

functional (adj)

funding (n)

fuse (n)

fuselage (n)

G

gain height (v)
gap (n)
gas turbine (n)
gasket (n)
gauge (n)
gear (n)
gear pump (n)
gearbox (n)
generator (n)
geometry (n)
gigabyte (n)
give off (v)
gold (n)
GPS [Global Positioning
   System] (n)
grab (v)
gradient (n)
grain size [of metals] (n)
gravity (n)
grease (n)
grid (n)
grille (n)
grip (v)
grommet (n)
groove (n)
ground support (n)
grounded (adj)
groundloop (n, v)
guard (n)
gyro-compass (n)
gyroscope (n)
gyro-system (n)

H

hacksaw (n)
halon (n)
hammer (n, v)
hand tool (n)
handle (n)
hand-operated (adj)
hand-propping (n)
handwheel (n)
hangar (n)
hardness (n)
hardware (n)
harm (n, v)

harness (n)
hazard (n)
headstock (n)
heat exchanger (n)
heat sink (n)
heavy-duty (adj)
HEI (high explosive
   incendiary) (adj)
height (n)
hertz (n)
high gear (n)
high-efficiency (adj)
hinged (adj)
hole (n)
hollow (adj)
horizontal (adj)
horsepower (n)
hose (n)
housing (n)
hover (v)
hum (n)
humidify (v)
humidity (n)
hydraulic (adj)
hydraulic power (n)
hydraulic system (n)
hydraulics (n)
hygrometer (n)
hygrostat (n)
hypoxia (n)

I

IAW (in accordance with)
icon (n)
identification code (n)
ignited gas (n)
ignition (n)
impact (n)
imperial (adj)
impression (n)
improvement (n)
in situ (adv)
inaccessible (adj)
inbound (adj)
incendiary round (n)
inch (n)

increase (n, v)
indicate (v)
indicator (n)
induce (v)
induction [stroke] (adj, n)
inductor (n)
industrial pollution (n)
inert (adj)
inflated (adj)
initiate (v)
injury (n)
inlet (adj, n)
in-line (adj)
innovative (adj)
inoperative (adj)
input (n)
insert (v)
inspect (v)
install (v)
installation (n)
instructor (n)
instrumentation (n)
insulate (v)
insulation sleeve (n)
intake (n)
integral (adj)
integral inlet (adj)
integrated circuit (n)
interchangeable (adj)
intercom (n)
interference (n)
intermediate chamber (n)
intermittent (adj)
intermix (v)
internal [structure] (n)
internal combustion
   engine (n)
interpret [data] (v)
interrupt (v)
interval (n)
inverter (n)
invest (v)
irrigation (n)
isolate (v)
isolator (n)

J

jack (n)
jaw (n)
jet fighter (n)
jet pump (n)
jetliner (n)
joystick (n)
junction box (n)

K

Kelvin (n)
kerosene (n)
key (n)
kilogram (n)
kilometre (n)
kilometres per hour (n)
kilowatt (n)
kinetic energy (n)
kit (n)
knob (n)
know-how (n)

L

L/H (left-hand)
lacquer (n)
lamp (n)
land selector (n)
landing gear (n)
landing gear leg (n)
landmark (n)
lateral (adj)
lateral axis (n)
lathe (n)
latitude (n)
layer (n)
layout drawing (n)
LCD television (n)
lead (n)
lead screw (n)
lead-acid [battery] (adj, n)
leak (n, v)
leg (n)
length (n)
lengthwise (n)
lethal (adj)
level (n)
lever (n)
life cycle (adj, n)

life jacket (n)

lift (n, v)

lifting (n)

lightness (n)

lightning (n)

lightweight (adj)

limb-restraint cord (n)

line [= tube] (n)

line up (v)

linear (adj)

liner (n)

lining (n)

linkage (n)

liquefy (v)

liquid (n)

lithium-ion [battery] (adj, n)

live (adj)

load (n)

loading (n)

lobe (n)

lock (n)

locking pin (n)

log (n)

longeron (n)

longitude (n)

longitudinal (adj)

longitudinal axis (n)

long-range (adj)

loop (n)

loose (adj)

loosen (n)

loss of load (n)

low gear (n)

LOX converter (n)

lubrication (n)

lubricity (n)

lug (n)

M

machine (v)

machine tool (n)

machinist (n)

magnetic field (n)

magneto (n)

magnifying glass (n)

main assembly (n)

main body (n)

mains (adj, n)

maintain (v)

maintenance (n)

maintenance check (n)

maintenance form (n)

maintenance manual (n)

malfunction (n, v)

malleability (n)

malleable (adj)

mandatory (adj)

manganese (n)

man-hour (n)

manoeuvre (n, v)

manual (adj, n)

manual procedure (n)

manufacturing (n)

manufacturing costs (n)

manufacturing industries (n)

margin of error (n)

mark out (v)

mass (n)

mass-production (n)

materials (n)

maximum (adj, n)

measure (v)

measuring humidity (n)

mechanical (n)

mechanical advantage (n)

mechanism (n)

medium-range (adj)

megaphone (n)

melt (v)

melting point (n)

memory (n)

message traffic (n)

metre (n)

metric (adj)

microscopic particle (n)

microswitch (n)

miles per hour (n)

military (adj)

milling machine (n)

minimise (v)

minus

misalignment (n)

mishandle (v)

miss [the problem] (v)

missile (n)

mix (v)

mixture (n)

mobility (n)

model (n)

modification (n)

moist (adj)

moisture (n)

molecule (n)

molten (adj)

momentum (n)

monitor (n)

monocoque (adj)

Morse code (n)

moulded plastic (n)

mount (v)

MP3 player (n)

MRO (Maintenance
   Repair and Overhaul) (n)

multi-cored (adj)

multi-graph (n)

multi-grip (adj)

multiple switch (n)

multiplied (v)

multi-purpose (adj)

multistage compressor (n)

multi-tubed (adj)

N

navigation (n)

navigation chart (n)

navigator (n)

NDI (non-destructive
   inspection) (n)

needle (n)

net [heat] (adj)

nickel-cadmium [battery]
   (adj, n)

nickel-metal-hydride (adj, n)

nitrogen (n)

non-level (adj)

non-metallic (adj)

non-precision aid (n)

non-structural part (n)

nose wheel (n)

notice (n)

nozzle (n)

nut (n)

O

OBOGS (onboard oxygen
   generating system) (n)

occupy (v)

ohm (n)

oil cooler (n)

oil head (n)

on-board (adj)

operate (v)

operating manual (n)

operation (n)

optics (n)

optimum (adj)

orientation (n)

orifice (n)

o-ring (n)

orthographic projection (n)

oscillator (n)

oscilloscope (n)

out of alignment (adj)

outbound (adj)

outflow valve (n)

outlet (adj, n)

output (n)

overall (adj, adv)

overhaul (v)

overhead [outlet] (adj)

overloaded (adj)

oxide (n)

oxygen mask (n)

ozone layer (n)

P

panel (n)

parachute (n)

parallel (adj)

part (n)

partially/fully extended (adj)

particle (n)

Pascal (n)

passenger (n)

passenger cabin (n)

path (n)

pattern (n)

payload (n)

PCB (printed circuit board) (n)

pedal (n)

percentage capacity (n)

perform (v)

performance (n)

permeability (n)

perpendicular (adj)

personnel (adj, n)

perspective (n)

phase (n)

physical (adj)

pictorial (adj)

pillar (n)

pin (n)

pinhead (n)

pipe (n)

piping (n)

piston (n)

piston head chamber (n)

pitch (n, v)

pitched (adj)

pitted (adj)

plate (n)

plated metal alloy (n)

platform (n)

play (n)

pliers (n)

plug (n)

plus (v)

pneumatic (adj)

pneumatic piston assembly (n)

point of view (n)

pointer (n)

pole (n)

polish (v)

pool (n)

port (adj)

potential energy (n)

potential freight load (n)

pound (n, v)

powder (n)

power (n)

power plant (n)

power rating (n)

power socket (n)

power stroke (n)

power tool (n)

power-to-weight ratio (n)

PPE (n)

precise (adj)

precision (n)

pre-drilled hole (n)

pre-flight check (n)

pre-flight test (n)

pressure (n)

pressure transducer (n)

pressurised fluid (n)

prevent (v)

price (n)

primary (adj)

primary flow (n)

prime (v)

priority (n)

probe (n)

process (n)

process water service (n)

processor (n)

production cell (n)

production line (n)

production site (n)

programming (n)

prohibit (v)

propane (n)

propel (v)

propeller (n)

propeller shaft (n)

property (n)

proportional to (adv)

protection (n)

prototype (n)

protrude (v)

provide [power] (v)

psi (pounds per square inch) (n)

pull up (v)

pulley [wheel] (n)

pump (n, v)

punch (n, v)

purge (v)

pushrod (n)

put back (in, together, etc.) (v)

put out [a fire] (v)

quadrant (n)

qualification (n)

quality assurance (n)

R/H (right-hand)

radar (n)

RADAR [Radio Detection and Ranging] (n)

radar tracking (n)

radial (adj)

radiator (n)

radio static (n)

radio-based system (n)

radius (n)

ram (n, v)

range (n)

ratio (n)

razor-sharp (adj)

react [with chemicals] (v)

reading (n)

reassemble (v)

reattach (v)

rebuild (v)

recalibrate (v)

recharge (v)

reciprocal lathe (n)

reciprocating piston (n)

reciprocation (n)

recirculate (v)

recoil (v)

record (v)

redirect (v)

redo (v)

reduction gearbox (n)

refine (v)

refit (v)

refrigerant (n)

refueling probe (n)

regulate (v)

regulation (n)

regulatory (adj)

reheater (n)

reinforce (v)

reinstall (v)

relative humidity (n)

relative to (v)

relay (n, v)

release valve (n)

reliability (n)

relieve [someone] (v)

remote (adj)

remote control unit (n)

remote switch (n)

remove (v)

repair (n, v)

repeatability (n)

replace (v)

replenish (v)

report (v)

reposition (v)

research and development (n)

reservoir (n)

reset (v)

residue (n)

resist (v)

resistance (n)

resistor (n)

restore (v)

restraint (n)

restrictor (n)

retract (v)

retract chamber (n)

retractable (adj)

reverse (adj, n, v)

revolutions per minute (n)

rheostat (n)

rib (n)

right angles (adj)

rigid (adj)

ring (n)

risk (n, v)

risky (adj)

rivet (n)

riveting tool (n)

robot technology (n)

robotic assembly (n)

rocker (n)

rocker arm (n)

rocket firing unit (n)

rocket motor (n)

rod (n)

roll (n, v)

rotary engine (n)

rotate (v)

rotation (n)

rotor (n)

rough (adj)

rough ground (n)

rubber band (n)

rudder (n)

run down [battery] (v)

runway (n)

S

saddle (n)

saddle washer (n)

safety (n)

safety equipment (n)

safety inspector (n)

sand (v)

sandblast (v)

sandpaper (n)

satellite (n)

saturated (adj)

saturation point (n)

saw (n, v)

scald (n, v)

scale (n)

schedule (n, v)

schematic (n)

scrapped (adj)

scratch (n, v)

screw (n, v)

screwdriver (n)

scribe (v)

scrutinize (n)

seal (n, v)

seat belt (n)

secondary (adj)

sector-shaped (adj)

secure (adj)

security (n)

select (v)

selector (n)

selector switch (n)

self-discharge rate (n)

semiconductor (n)

semi-monocoque (adj)

sensor (n)

separator (n)

service tag (n)

servicing equipment (n)

set (v)

set up (v)

shaft (n)

shallow angle (n)

sharp (adj)

sharpen (v)

shear (adj)

shearing strength (n)

shears (n)

sheath (n)

sheet (n)

shift (n)

shock (n)

short circuit (n, v)

shorted (adj)

short-range (adj)

shutdown (adj)

sick bag (n)

side cutters (n)

sidewall (n)

sideways (adj, adv)

sign (n)

sign off (= authorise) (v)

simple transformer (n)

simplicity (n)

simultaneous (adj)

single pole switch (n)

single-skin construction (n)

site (n)

size (n)

skater (n)

skeleton (n)

sketch (n)

skid (v)

skin (n, v)

slack (adj, n)

sleeve (n)

slip ring (n)

slope (n)

slosh (v)

slow down (v)

smith (n)

smooth (adj)

snag (v)

socket set (n)

software (n)

solder (n, v)

solder sucker (n)

soldering gun (n)

soldering iron (n)

solenoid (n)

solid (adj)

solid state (adj)

solidify (v)

sophisticated (adj)

source (n)

spar (n)

spare part (n)

spark (n)

spark plug (n)

specific heat capacity (n)

specific operating
   conditions (n)

specification (n)

speed (n)

sphere (n)

spin round (v)

spindle (n)

splice (v)

sponsor (n, v)

spot [the problem] (v)

spray (n, v)

spread (v)

spring (n)

spring-loaded (adj)

square (adj, n)

square root (n)

stabilise (v)

stable (adj)

stainless steel (n)

stall (v)

standard rivet (n)

standard specification (n)

starboard (adj)

starter (n)

starter motor (n)

starter-generator (n)

state (v)

static discharge (n)

static line (n)

stationary (adj)

statistics (n)

steady (v)

steering (n)

stiff (adj)

stiffening (n)

stitch/stitching (n)

storage (n)

straight line (n)

strain (n, v)

strainer (n)

strap (n)

streamlined (adj)

strength (n)

stress (n)

stressed skin panel (n)

stringer (n)

strip down (v)

stroke (n)

structural part (n)

structure (n)

strut (n)

sub-assembly (n)

suck [in] (v)

suction (n)

suffocate (v)

suitability (n)

suitable (adj)

sulphur (n)

sump (n)

supercharger (n)

supply (n, v)

support (n, v)

surface (n)

surface area (n)

survivability (n)

survival (n)

switch on/off (v)

synthetic (adj)

systemic [problems] (adj)

systems engineer (n)

tags [warning ~] (n)

tail (n)

taildragger (n)

tailor (v)

take apart (v)

take out [of service] (v)

take over (n, v)

take-off (n)

take-off selector (n)

tangled (adj)

tank (n)

tape (n)

tape measure (n)

tarmac (n)

tattle tale (n)

taxi off (v)

tear (n, v)

technician (n)

technique (n)

technological (adj)

technology (n)

teeth (n)

telescopic (adj)

temperature reading (n)

tensile strength (n)

tension (n)

terminal (n)

terminal block (n)

terminology (n)

thermal conductivity (n)

thermostat (n)

thickness (n)

thread (n)

thread pitch (n)

threat (n)

three-dimensional (adj)

thrust (n)

tie (v)

tighten (v)

timing chain/belt (n)

tin snip (n)

titanium (n)

tolerance (n)

top (n)

torque (n)

torsion (n)

toughness (n)

tower [radio] (n)

towing (n)

trace [fault] (v)

trainer (n)

transfer (n, v)

transistorised static
    inverter (n)

transmit (v)

transmitter receiver (n)

transparent (adj)

transverse axis (n)

trap (v)

tread (n)

treadle (n)

trial (n, v)

trickle-charge (v)

tricycle (n)

trim (n, v)

troubleshoot (v)

troubleshooting (n)

trussed (adj)

tube (n)

turbine (n)

turbofan engine (n)

turboprop (n)

turbulence (n)

turn around (v)

turn over [engine] (v)

turn-coordinator (n)

turning tool (n)

twist (v)

twisted (adj)

two-/three-point
    perspective (n)

two-dimensional (adj)

two-way radio (n)

tyre (n)

ullage (n)

undercarriage (n)

undertake [maintenance] (v)

uneven (adj)

unit (n)

update (n)

upwards (adv)

vacuum (n)

value (n)

valve (n)

vane (n)

vanishing point (n)

vaporise (v)

vapour (n)

velocity (n)

vent (adj, n)

ventilate (v)

venturi principle (n)

verify (v)

versatile (adj)

vertical (adj)

vertical axis (n)

vertical take-off (adj, n)

vibration (n)

vice (n)

visibility (n)

visualise (v)

void (n)

volatility (n)

volt (n)

voltage (n)

voltage probe (n)

voltage regulator (n)

voltmeter (n)

volume (n)

VOR [Very high-frequency
    Omni-directional Radio
    range] (n)

vulnerability (n)

war plane (n)

warning (adj, n)

washer (n)

waste (v)

watt (n)

WD40 lubricant (n)

weapons system (n)

wear (v)

weight (n)

weld (v)

welding (n)

well-ventilated (adj)

width (n)

wing (n)

wing box (n)

wingspan (n)

wire (n)

wire strippers (n)

withstand (v)

wood turning (n)

work cage (n)

work order (n)

work table (n)

workpiece (n)

works [= maintenance] (n)

wrap around (v)

wrench (n)

wrought iron (n)

yaw (n, v)

zinc (n)

Tapescript

Unit 1

Unit 1, Lesson 1, Track 1

1.

A: So this is your new telly then?

B: Yep – what do you think?

A: Brilliant picture … and it's so thin.

B: And it's really light … only about 18 kilos … I can lift it easily!

2.

A: Have you ever been on a hovercraft?

B: Yes. We went across to France on one last year.

A: What was it like?

B: Amazing! I didn't know if I was on a boat or a plane. We were doing nearly a hundred and twenty kliks and we were only about a metre above the water. Fantastic!

3.

A: Look, this is the one I'd recommend Bob. It's got a massive memory.

B: Really?

A: Yeah, 60 gigabytes. You'll be able to play your horrible music for hours!

B: Ha ha.

4.

A: Whoa, what was that?

B: Oh, it's one of the Harriers from the air base along the coast. It's probably doing a thousand kilometres an hour.

5.

A: Can I help you, sir?

B: Yes. I'm interested in this machine here. Can you tell me something about it?

A: Sure … Well, of course, there's no bag to worry about and it's got a big strong motor, 1,400 watts in fact … And a 2 litre bin … So you don't have to empty it too often.

6.

A: That's the new Mazda RX 8, isn't it?

B: Yeah. It's a lovely motor. 230 brake horse power. That's far more than you usually get from smallish cars like that. I bet it goes like a rocket on the motorway!

Unit 1, Lesson 4, Track 2

Host: Good evening and welcome to amazing animals. This evening, we are going to hear about one of nature's strongest materials, made by some of nature's smallest creatures, the spiders. To tell us more about these remarkable little insects is Dr Donald Parsons, the director of Hopewell Zoo … Good evening, Doctor Parsons.

Dr P: Good evening, Robin and … er … before we go any further can I er just correct one small mistake …

Host: Oh … yes, of course.

Dr P: Yes … well, you see the thing is, spiders are not actually insects.

Host: Oh really?

Dr P: No, they actually belong to a group of creatures called arachnids, which have eight legs, not six, like insects and they are more related to crabs and scorpions than they are to flies and beetles.

Host: Oh, well, thank you for putting me straight about that … Now, I believe that you have got some interesting information for us about the silk that spiders use to make their webs with.

Dr P: Yes, that's right. And not just to make webs either. They use the silk to wrap up the small creatures they catch in their webs, to make shelters where they can hide from their enemies and even as lifelines to help them escape when they are being chased.

Host: So this spider silk must be a pretty amazing

material. I know I get spider webs on my car wing mirrors in the morning and they don't blow off even when I'm driving fast.

Dr P: That's right. The silk often contains a strong adhesive, it's very, very sticky. Like chewing gum … but also extremely, extremely strong. Did you know that someone once calculated that a length of spider silk only the thickness of a pencil could stop a Boeing 747 airliner, without breaking.

Host: Incredible!

Dr P: Yes, weight for weight, it's much stronger than steel … up to 5 times as strong under the right conditions. If I can give you a few technical details …

Host: Well I don't …

Dr P: Yes there's something called Young's modulus of elasticity, which is a measurement of the tensile strength, how much you can stretch something before it breaks really … and this is measured in units called Pascals. The radial thread of a spider's web …

Host: Radial thread?

Dr P: Yes … the line that goes from the centre of the web to the edge.

Host: Uh huh …

Dr P: Yes, well that radial thread has a tensile strength of well over a thousand Pascals, whereas mild steel, by contrast comes out at about 400 Pascals.

Host: Hmm. That's a big difference. Does it have any other interesting properties?

Dr P: It certainly does. It is very, very light stuff, 25% lighter than synthetic plastics made from oil. And it keeps its strength at very low temperatures – down to minus 40 in fact. At that temperature, a lot of materials become very brittle and really quite useless. It is also fairly resistant to moisture, more so than the silk that is used to make clothes …

Host: But if it's so marvellous, why don't we see it being used more often? It sounds like an almost perfect material.

Dr P: Well, the problem is, it's almost impossible to collect it in any useful quantities. The silk is very, very fine – finer than a human hair and you need an awful lot of it to make a useable amount. And

you can't farm spiders like silkworms. They don't do as they are told. But seriously, spider silk has occasionally been used in the past, to make the cross hairs in instruments and gun sights, for example. However, there may be a chance of producing more of it in the future.

Host: Really?

Dr P: Yes, apparently, some biologists and chemists in Canada are trying to produce a type of spider silk from goat's milk.

Host: Goat's milk?

Dr P: Yes, it's really most interesting. Their idea is to mix the genes of the …

Unit 2

Unit 2, Lesson 2, Track 3

PM: Welcome to another edition of collector's corner. I'm Phillip Martin. In today's programme, we'll meet someone who claims to have the world's largest collection of bicycles, someone with a wonderful collection of Arab coffee pots and lastly a chap who collects … wait for this … you're not going to believe it … the labels from tea bags - yes, tea bag labels are now collectable! You heard it here first.

But to start today's programme we're going to talk to Richard Bolton about that traditional weapon of the Japanese warrior, the Katana sword.

Hello, Richard and welcome to the programme. Now you've been collecting Japanese swords for quite a number of years I believe?

RB: Yes, that's right. It all started when I went to Japan on a business trip about 15 years ago. I was lucky enough to be taken to see a Japanese swordsmith at work while I was there. I found it so fascinating that I became completely hooked … and … well I've just been collecting Japanese armour and weapons ever since.

PM: How many swords have you got altogether?

RB: Seven.

PM: Is that all ... forgive me ... uh ... it's just that it doesn't seem like very many.

RB: I know ... but you have to remember that it's quality, not quantity that counts! These weapons are very, very expensive and each one has been produced by a different master craftsman ... and a couple of them are well over a hundred years old ... But each one is special and different in its own way.

PM: Can you tell us a little about how they are made?

RB: Well in two words really ... heating and beating.

PM: Is that it ... surely not?

RB: Well, I am making it a bit simple ... but you see ... the uh ... technique developed because originally the Japanese were only able to produce rather impure metal.

PM: Why was that?

RB: Because Japan had very little high quality coal so it was difficult to heat metal to very high temperatures in the furnaces.

PM: Right ... so the work that might have been done by the heat had to be done by the craftsmen.

RB: Yes that's it. Basically the sword maker repeatedly heats, folds and hammers the metal. This repeated folding and beating eliminates any air bubbles or gaps in the metal and makes sure that the carbon content of the steel is uniform - it homogenised it ... and the continual heating, cooling and re-heating helped get rid of ... to eliminate many of the impurities in the rather low-quality steel.

PM: And what is the end result of all this work? What kind of sword do you get at the end of it all?

RB: Quite an unusual one actually. For a start, it's rigid, not very flexible ... designed to be used edge on like a guillotine. And there is only one sharp-cutting edge, not two ...

PM: Ugh ... don't go on ... it sounds nasty.

RB: Yes ... sure ... but did you know that the techniques used in making swords were also used to make gardening and woodworking tools, planes, chisels, cutters, knives and so on. A lot of ...

PM: Really ... how interesting ... and presumably there was some special armour made to protect soldiers from these deadly blades?

RB: Armour ... oh yes, that's another subject in itself ... you know most people ... think of heavy metal when they hear the word 'armour' ... but in fact, the Japanese armour consisted mainly of bamboo wood, leather and quilted cotton which was ....

Unit 2, Lesson 4, Track 4

1. The drill bit can be moved in more than one axis. (A)
2. The handle is usually made of plastic. (A)
3. Power is transmitted to the drill bit by the gears. (BOTH)
4. Its wide heavy base helps to prevent unwanted movement. (B)
5. The pillar supports the work table. (B)
6. The drill is powered by compressed air. (A)
7. It's unsuitable for drilling teeth. (B)
8. The drill bit can be accurately positioned. (BOTH)
9. Many people are afraid of it. (A)
10. The position of the bit can be controlled by the handwheel. (B)

Unit 2, Lesson 6, Track 5

1. Before starting the machine, ensure that the feed engage lever and the thread cutting lever are disengaged.
2. Select either longitudinal or transverse feed axis by means of the push-pull knob on the apron.
3. Use the small selector handle at the bottom of the gearbox to determine the direction of feed.
4. Select the feed rate required by setting the selector dial and the three feed-selector handles at the top of the head-stock.
5. Use the two handles at the top of the gearbox to select the spindle speed.
6. Switch on the electrical supply at the mains isolator.
7. Start the spindle by raising it for forward rotation or lowering it for reverse.
8. Start and stop the feed motion as required, by means of the feed-engage lever.

9.   Always stop the machine before changing speeds or feeds.

Unit 2, Lesson 6, Track 6

The lead screw is a long, threaded rod that carries a tool along the axis of a rotating workpiece. It ensures that the workpiece moves at a constant, even speed so that threads can be cut into it. The relationship between the longitudinal speed of the tool and the rotational speed of the workpiece can be varied by means of a gearbox.

Unit 2, Lesson 8, Track 7

T: John, Martin can you come over here ... ok ... good ... now have you done ... what I asked?

J: Yes, we've cut two square sheets of this metal you gave us, hang on, here you are, is that all right?

T: Hmm, it looks ok but, uh ... Martin, run your thumb along the edge there, yes, ... that one.

M: Ouch, it's a bit sharp.

T: Yes, I thought so ... it's really important to smooth off the edge of any metal you cut. You know why?

J: Yes, cos somebody could get hurt.

T: That's right, safety is always important, but there's another reason to do with this particular metal.

M: What, you mean aluminium sheet?

T: Well this isn't ordinary aluminium, though, it's an alloy, which has been through a process called alcladding. It's had a skin of pure aluminium applied to it on both sides.

M: What's the point of that?

T: It's to stop corrosion ... The aluminium forms a protective oxide skin. But this means that you have to handle it and work with it very carefully because ...

J: ... you might break the skin and then the metal underneath ... you could get corrosion in underneath the skin.

T: Exactly ... and in an aeroplane, thousands of feet up in the sky, that's dangerous ... very dangerous.

M: Does this stuff have a special name?

T: This particular sheet is called 2024-T3 ... It's used a lot in aircraft.

J: So how exactly do you have to be careful when you work with it?

T: Well, think for a moment ... What are the four main operations you do on sheet metal?

J: ... er ... cutting, drilling, bending ... er ...

M: ... and marking out ...

T: ... exactly, ... marking out ... Before you do anything else ... and that's when you have to start being careful ... from the beginning.

M: ...when you're scribing the lines.

T: Ah ... but that's the first thing to remember. You mustn't use a scriber to make the lines and points. Scribers can cut too much into the oxide surface, which allows corrosion in ... Also, ... where you cut into the metal, it makes it a little weaker along those lines ... and if ... there is vibration and stress in the metal ... it can fracture there. It's a bit like breaking the pieces off a bar of chocolate. The chocolate breaks along the lines marked in it. So you should always use a special marker pen like this ... It's called a sharpie ... you can get them in different colours, but I prefer blue, it seems to show up better.

J: What about cutting the sheet? Are there any special things to remember?

T: Yes, three things: finishing, finishing and finishing. You must make sure that any cut edges are rounded off and completely smooth. Rough and jagged edges are like small cracks in the metal ... And these can get bigger and bigger. The best test is to check that you can't see the marks of the saw teeth ... or the guillotine blade if you use the treadle shears. The same thing when you drill a hole ... you must remove all the burrs and make sure the inside of the hole is completely smooth.

M: And what about making bends?

T: Now that's quite a complicated subject ... But the basic rule is that you mustn't put too much stress on the metal. You mustn't bend it over too far or with a bend radius that is too small for the metal you are working with.

M: How do you know if you've done that?

T: You make sure you don't by looking at the special tables ... no ... no, I think that's enough talking for today. We'll leave it until tomorrow. What I want you two to do now is mark out these two sheets, and drill the holes here and here, just like you can see on this drawing ... you see ... here and here ... and then bring them back to me. And remember you must remove all the burrs and smooth all the sharp edges. You should be able to finish it in about half an hour. But don't worry if you can't. You don't have to finish it today. The important thing is to do it correctly ...

J &M: Right ... Ok ...

T: You can use that bench over there.

Unit 2, Lesson 10, Track 8

I: So, Robert, this is the plane you built yourself?

R: Yes, this is my baby, what do you think of her?

I: Well, to be honest, I'm amazed! It looks just like a new plane from a factory. It's so well finished. You must have done a lot of studying and technical training before you started building it!

R: No, not really. Of course I've always been interested in planes and I've had a pilot's license for over ten years now. But my normal job is in a bank ... and ... apart from a few household repairs, I don't have much hands-on technical experience – At least I didn't before I started building her.

I: And how long did it take you from start to finish?

R: Hmm ... let's see now ... this is August and the kit was delivered in April the year before last ... yes, so just over two years.

I: And how much of that time did you spend on the project?

R: Practically every weekend really, as well as a lot of evenings after work. I'm afraid my family didn't see very much of me. I suppose it was about 800 hours altogether.

I: You said it was a kit?

R: Oh yes, I couldn't have built it from scratch. A lot of the difficult stuff had already been done at the factory.

I: And where did you actually make it?

R: In my garage.

I: Your garage?

R: Yes. Of course, I had to move the car out and put a work bench in!

I: I bet you did! You must have a pretty big garage.

R: Hmm, biggish but not enormous. What is it ... Yes ... It's about 24 feet long, 9 feet wide and nine feet high. Of course, I had to fit the wings on outside. But I did most of the construction inside.

I: You must have needed a lot of special equipment.

R: No, not really, although I did buy myself a kit of good hand tools.

I: Such as?

R: Oh, you know, the usual thing, screwdrivers, pliers, saws, spanners and so on. Oh yes, and a pair of tin snips for cutting the sheet metal.

I: So just those hand tools then.

R: Not exactly. I did get myself a new electric drill ... the old one was on its last legs. And a hand rivet gun for the blind rivets.

I: 'Blind' rivets?

R: Yes, sorry, it just means a kind of rivet that one person can fit easily on their own into any part of the plane. You only need to work from one side.

I: Right ... and tell me ... how much did all this cost, apart from your time, of course?

R: Just under fourteen thousand pounds for the kit, plus about two hundred for the tools. Of course, I did have to take my family away for a special holiday to make up for all the time they didn't see me. That was a couple of thousand for all of us!

I: And one last question, if you don't mind.

R: Sure.

I: Why did you do it?

R: Ah. Well, partly the cost, it was a lot cheaper than buying one ready-made. But really, I suppose it was also because I enjoy a challenge. And I wanted to do something that was completely different to my work. You know, even if I never build anything else I'll always be proud of this … it's given me a lot of satisfaction.

Unit 3

Unit 3, Lesson 3, Track 9

L: No, it's no good … I can't see it. My eyes just aren't as good as yours. There's not enough light for me yet.

G: Don't worry. Be patient. Close your eyes and then look across the lake again. He's on a branch about halfway up that very tall tree on the left of the rocks. Start at the bottom of the tree, and move your eye slowly up until …

L: Ah yes, there he is! My gosh, look at him … if he just stays still long enough while I get my binoculars … There, yes, I can see him clearly … every feather. He's a big chap.

G: Yes, fully grown … and he's going to start hunting in a minute, I think. Don't use your binoculars; you won't be able to follow him. He's much too fast. Do you see those circles in the water over to your left? That's a big fish just coming up to feed on the early morning insects. I'm sure the eagle must have seen him and he'll … yes there he goes.

Unit 3, Lesson 3, Track 10

L: There he goes, he's taken off from the branch. Will you look at those wings - the wingspan must be at least a metre. He's not going for the fish though, he's climbing away. What's he doing? He's a couple of hundred metres beyond where the fish are feeding. Ah, now he's turning back, banking towards us.

G: Yes … he's got the height he needed to start his descent. He'll dive at quite a shallow angle so the fish won't see him coming …

L: … and what a speed, too! He must be doing at least 60 km an hour.

L: Got it! … Wow, he just picked the fish up as if he was in a supermarket … wait a second, he seems to be having some trouble - he's just dragging it along the surface of the water.

G: Yes, he can't … the fish is too big and heavy. He can't produce enough lift from his wings … no, I

think he's ok. He's beginning to climb. But he's certainly a lot slower … Oh, I think you are a lucky man – quickly, look above him and to the right … you see?

L: Woooh! It's another eagle - it's diving straight down towards him. What's it doing?

G: Wait, you'll see. It's a female … they are usually bigger and heavier.

L: I don't believe it! Is it trying to steal the fish?

G: Yes. If the first bird drops it, the other one will dive down and grab it in mid air. They are famous thieves, these eagles. But it's unusual to see one fish eagle steal from another; they usually attack other birds.

L: What a sight! They've both got their talons into the fish and they're flapping away … he really doesn't want to let go of his breakfast. Is she trying to pull him up higher? She seems to be. Why's that?

G: Yes. She is. If she can gain enough height, she will try a special manoeuvre … yes … watch.

L: Ah, look at that, she's rolling and falling at the same time, and the male is having real trouble keeping up.

G: Yes. Because the female is bigger and stronger, she will try to twist the fish away from him or make him drop it, which he may have to do in the end to avoid falling into the lake. There … he's dropped it.

L: Woooah … and she's shot down past him and grabbed it … ha! … and she's off with his fish. She's climbing really fast … she must have some power in those wings. Fantastic. She's away. Well, what a lot of excitement - and the day has only just started! I wonder what we'll see next.

G: Well it's your lucky day for sure. Perhaps a lion will come down to drink … there's still time.

Unit 3, Lesson 6, Track 11

The final production assembly line for the German version of the Eurofighter is located in the South German town of Manching. It is here that, as well as equipping the main fuselage, the engineers and technicians put together more than 300 pieces of equipment, sub-assemblies and assemblies to produce the finished aircraft.

At the first stage, the three main assemblies are fitted or spliced together. These are the centre fuselage, the rear fuselage and the cockpit. Next, the flight control surfaces such as the wings, flaps and fins are attached to the fuselage. At this point, the fighter starts to look like a real plane. At the third stage, all the electrical cables are thoroughly tested and then the aircraft is moved on to Station 4, where the mechanical, electrical and hydraulic systems are subjected to rigorous testing with detached computerized equipment. Following satisfactory completion of these system tests, the aircraft is ready to have its engines and weapons systems fitted.

Following this, the flight control and navigation systems are installed and the plane is now ready for its pre-flight tests. Once these have been carried out, the aircraft is tested in flight, before finally being moved to the paint shop to be painted in the colours of the German air force.

Unit 3, Lesson 9, Track 12

A: I left university eighteen months ago with a degree in maths and physics and an MSc in metallurgy. Since then, I've been working for a testing laboratory in London. It's quite a small firm so we all have to deal directly with customers face to face, especially if we have an urgent job on ... which is quite often! We've recently had quite a few jobs for one of the big aerospace companies, mainly on stress fractures as well as routine structural loading tests. That's how I became interested specifically in the aircraft industry. I know I'll have to start at the bottom and work my way up. That's no problem. And I'm not married yet, so I don't really mind where I work.

B: I've lived in the UK all my life. All my family's here. I'd never want to move away. I originally wanted to go to art school. I was always drawing and painting as a kid. I still do in my spare time. Thing is, there's no money in it ... so anyway, I took my Dad's advice and got an engineering degree. Luckily, I was good at maths and science, too! My job combines both, art and engineering. Mind you, I'm having to use computer aided design more and more these days. I don't stay long with any one company. I prefer short term contracts ... I think the longest I've done is eight months. I work away, sometimes. You know ... find a comfy bed and breakfast for Monday to Friday and then head back home at the weekends.

C: I was in the airforce for 20 years after I finished my apprenticeship ... had a wonderful time and went all over the place ... Germany ... Cyprus ... Malta. I worked on helicopters and Hercules transports mainly, skin fabrication mainly. That's my speciality, I suppose you could say. Since I came out of the air force, I've been working locally as a machinist, but to tell the truth, I'm getting a bit bored. It's the same thing day in, day out and really I'm a sheet metal basher at heart. I'd like to go somewhere sunny for a few years, preferably the Middle East, and try something new. I'd be able to save a bit more towards my pension too!

D: Mathematics is my chosen field. I've always loved it. I did a degree in maths and physics and then an MSc in avionics and control systems. I've been working on a Doctorate and teaching at the University at the same time, I've also done quite a few short contract jobs for the air force. I'm looking for something long term, a proper career in the systems and navigation field. I'd really like to work on a project from the early stages right through to completion. My family is from Italy originally. They came to South Wales to work in the coal mines during the last century. But I love it here – such a beautiful place – I would never move away.

Unit 4

Unit 4, Lesson 2, Track 13

T: Right ... ok ... if I could have your attention, please. Good. Now, before the break, we were talking about various different ways that farmers could pump water on to their land. We looked at very hot dry countries where there isn't much water anyway. Now I want us to study a device which can be used when you don't have a shortage of water, but when the problem is that the water is in the wrong place. Can anybody think of a situation like that? Yes, John.

J: Well ... in countries where they get a lot of heavy rain, one part of the year and then it goes very dry. You might want to fill up a reservoir or some storage tanks ... so you have water all the year round.

T: Exactly. Good. Yes, Peter?

P: Or maybe you need to supply animals with drinking water in fields which are high up ... or water crops maybe?

T: Yes, that first point is a good one. Any other ideas? No. OK, right. Now I'll just put this diagram up on the board. There. Now, the basic idea is this. You see here where the water comes down through the pipe at point A. This is the primary flow, water flowing through a pipe on a slope – so, what force is making the water flow, anybody?

Student: Gravity.

T: Excellent! So gravity is our source of energy for this machine. And of course, gravity is free! Now, the water flows down through the pipe towards this valve at the bottom. And what do you think will happen as it hits the valve?

J: It will force the valve closed.

T: Yes good - and this is why it's called a ram; because it pushes, or rams, the valve shut. But now - and this is the really clever bit - because the moving water has nowhere to go, there is a sudden increase in pressure. This is called 'water hammer'. You get a reaction in the opposite direction to the flow. To help you understand this, think about what happens if you throw a stone at a wall. The stone is moving and then it is suddenly stopped. Does it just drop down vertically?

Various students: No/it bounces/it comes back/it rebounds.

T: Exactly. So when this happens to the water, there is this backwards pressure in the pump, and this is enough to open the delivery valve here in the centre of the diagram, and push water upwards through the outlet pipe. Of course, then the pressure inside the pump goes down again, and as it returns to normal, our spring loaded valve at the bottom - it's sometimes called a 'clack' valve because it makes a clacking sound like someone hitting two pieces of metal together ... actually, sometimes it doesn't have a spring, just a weight - anyway this valve opens again, and of course the – what happens to the delivery valve up here?

A: It, er, closes again, because the pressure inside the delivery pipe has dropped.

T: Well done. Yes, and so the whole cycle starts again: first, gravity moves the water – the moving water has momentum - which is what, anybody?

Various students: Mass multiplied by velocity.

T: That's right. Mass times velocity. And when the water hits the closed valve, this energy of momentum is changed into pressure - is changed into work. It lifts the water up the outlet pipe. So there you have it, a three part cycle: momentum, pressure, work; momentum, pressure, work, and so on, about 30 to 60 times a minute. As long as there's water coming down the pipe, the pump just keeps going and you can move the water up, to any other place you need it. It only has two moving parts and it doesn't need an electric motor or an engine to power it. Now, has anybody got any questions?

Unit 4, Lesson 5, Track 14

So when this happens to the water, there is this backwards pressure in the pump, and this is enough to open the delivery valve here in the centre of the diagram, and push water upwards through the outlet pipe. Of course, then the pressure inside the pump goes down again, and as it returns to normal, our spring loaded valve at the bottom - it's sometimes called a 'clack' valve because it makes a clacking sound like someone hitting two pieces of metal together … actually, sometimes it doesn't have a spring, just a weight - anyway this valve opens again, and of course the – what happens to the delivery valve up here?

Unit 4, Lesson 5, Track 15

… the use of hydraulics in matters of life or death. If I can take an example that everybody's familiar with, which is the domestic cat. Cats are good climbers and will often jump from a tree at a bird, or sometimes just slip, and fall. But, a fall that would kill or badly injure a human or another animal, may have little effect on a cat, because they have two characteristics which are essential for a safe landing, and which the aviation industry, among others, makes use of.

Firstly, cats have a small but extremely sensitive fluid-filled organ in their inner ear. When they move, so does the fluid, and so they know precisely their position relative to the ground at all times, and in a landing situation, of course, that's vital information. The equivalent in an aircraft is an instrument called the artificial horizon, which tells the pilot his orientation. As soon as a fall – or a descent, in the case of an aircraft – starts, this detector enables the animal, or pilot, to determine their orientation quickly so that they can prepare for the landing impact.

That's one important feature. Then, secondly, to help prepare the body itself for landing, the physical design needs to enable the falling body to easily get into the best position. Cats benefit from an extremely flexible skeleton. They don't have collar bones, and the bones in their backs have an especially high mobility. This enables them to twist and turn their bodies quickly and easily in an emergency landing, so that by the time they hit the ground they are oriented in the best position to absorb the shock. Aircraft don't have anything like that structural mobility, although there is some flexibility built into the airframe.

But what a pilot can do is to use the aircraft's control surfaces – the surfaces that move- like the rudder, flaps, etc. He can use them to slow the aircraft's descent – something that a cat can't do. This will minimise the shock of landing and give him time to manoeuvre. The control surfaces also allow him to … to manoeuvre into the best position very accurately.

The cat looks at the ground to help here; so, of course, does the pilot. This is where hydraulics comes in. To distribute the load, cats always land on all four feet simultaneously; aircraft, however, land just on their two main wheels or wheel groups. Cats curve their backs, which aircraft can't do of course, but where animals use muscles to minimise the shock of the impact, aircraft have to make use of hydraulics: as we saw with the air brake, a hydraulic chamber is filled with pressurised fluid.

When the aircraft hits the ground, the landing gear assembly basically acts as the piston inside the chamber, pushing against the fluid, which absorbs the force of the landing.

All being well, both the animal and the aircraft will land safely – although, hopefully, the aircraft will have many more than just nine lives …

Unit 4, Lesson 6, Track 16

J: Hello George, haven't seen you for ages … must be what … Seven or eight months. We thought you'd left the club but forgotten to say goodbye.

G: No … No … I've just been so busy at work … there just hasn't been enough time to fit the flying in … Anyway how are you, Jack … How's the family?

J: Fine … fine, thanks. We're all well. My eldest son's just gone off to university. And you and yours?

G: Yep … all fine … As I say, I haven't had much free time … but things are a bit quieter now so I'm looking forward to getting some more flying in.

J: I see you're still flying with that taildragger landing gear. You should really move with the times.

G: No, thanks. I'm quite happy with this layout, thanks. I know it's more difficult to take off and land but I enjoy the excitement and it keeps my flying skills up to scratch. Anyway there are some real advantages to this taildragger gear you know.

J: Oh yes, such as ...?

G: Well for a start, it's much easier to land on rough and uneven ground. There's no danger of breaking the nose wheel in a hole or digging up the field with the propeller when the nose tips forward. I'd much rather make an emergency landing in a field in my plane than in yours.

J: Hmmm ... that's true, I suppose. But what about on the runway ...? I mean the centre of gravity is behind the wheels so the moment you get out of line, the plane's going to start trying to spin round ... and ... the cockpit is pointing up at the sky. You can't see properly.

G: Oh you get used to the visibility problem, you just have to pay more attention. But I agree there is a danger of spinning round during take off and landing. In fact, I have done a couple of ground loops myself. But that was when I was a beginner. If you know your plane and pay attention to the wind speed and direction, it's not a problem.

J: But what about when you're parked on the runway? There's much more danger of the plane being damaged by sudden gusts of wind, with the wings angled upwards like that. You remember what happened when we had that bad storm a couple of years ago. David's plane was a write off.

G: Yes, that's true. I admit that can be a problem. But you just have to make sure you tie the machine down well or put it in the hangar.

J: And another thing ... I can get in and out of my plane far more easily than you can ... and it's much easier and quicker to load stuff in because everything is level with the ground.

G: Ah yes, but I can fly faster when I'm in the air because my landing gear is smaller. It weighs less and it doesn't have as much air resistance ... I

save quite a bit on fuel as well. If I had the tricycle gear fitted, I'd be spending about ten per cent more on fuel, I reckon.

J: Hmmm ... well, I can see we're never going to agree ... Brrr ... it's getting cold out here ... Let's get to the clubhouse and grab some hot coffee.

G: Now that I do agree with!

Unit 4, Lesson 8, Track 17

M: Hi, I'm back!

F: How did it go?

M: Fantastic! I actually got to fly the plane by myself!

W: You mean they let you go up alone?

M: No, of course not. That's against the law. Dennis - the instructor - was there all the time. But, I did actually steer the plane and made it go up and down. It's got dual controls.

B: What's that?

M: It means you have your own set of controls but the instructor can take over from you instantly if he needs to. Anyway I'm going to have another lesson as soon as I've saved enough money. It's just such a wonderful feeling up there!

F: Slow down, slow down. Why don't you tell us all about it from the beginning?

M: Sorry, yes, well I got to the airfield and Mr Saunders - Dennis - he began by showing me the main controls in the cockpit and explained what they do.

D: Now, before you get in, I'm just going to show you the main controls and how they relate to the flight control surfaces. Those are what we call the bits of the plane that move about. So if you stand just there while I get in ...

D: OK. First, here's the control column. That's the correct name for it, but it's sometimes called the joystick, or even just ... just the 'stick'. Now I can move it in the longitudinal axis, backwards and forwards ... and I can move it laterally, side to side. Now if I push it forward like this - look at the back of the plane. You can see the elevators moving up and down – so it also controls movement in the vertical axis.

M: Oh yes, I see.

D: Now, when I pull the stick back the elevators move up. And it's this control we use for what's known as the pitch of the plane - raising and lowering the front end, the nose.

M: Right. So you can use it to increase or decrease your flying height.

D: Yes that's it. Now, as I said, we can also move the stick from side to side. Look, watch what happens.

M: Ah yes, those flaps on the wings are moving up and down.

D: Well, actually they aren't flaps. The flaps are actually further in, near the fuselage, you see.

M: OK. What do they do?

D: Well, we mainly use those for take-off and landing. You won't be doing either of those things today, so don't worry about it for now. No, these control surfaces are called ailerons and they control the rolling movement of the plane, sometimes called banking. You see, if I want the plane to roll or bank to the left, I move the stick to the left. If I want to bank to the right, I have to move the stick laterally to the right.

M: OK, yes, I've got it.

D: And lastly, there are these two pedals down here. Now if I press them down with ...

M: Ah yes, it's like a ship's rudder.

D: Yep, that's it. It is the rudder. We use it to swing or turn the nose of the plane to the right or left. It's called yaw. Y-A-W: yaw.

M: So that's the control we have to use if we want to turn the plane round.

D: Well, in fact, not exactly - you mainly use the ailerons for that - but yes, you still need the rudder to help you control the turn.

M: How are the cockpit controls connected to the surfaces? Is it electrical or hydraulic?

D: In this little plane, neither! All the linkages are mechanical. It's all rods and cables, so you can really feel directly how the plane is behaving.

M: Right.

D: So. Now you know about the primary control system, I think that's enough theory for the moment. Now we're going to do all the pre-flight checks and then we're off.

M: Great! So what checks do you have to do?

Unit 4, Lesson 10, Track 18

Commentator: Hello and welcome to our radio listeners to the Spanlow Air show. As you can probably hear, there's a large, expectant crowd here today and we're all waiting for the opening event, which is of course, a twenty-minute display by the famous Red Arrows aerobatic team ... And I've just been told that we're about to start ... and yes, here they come ... All nine planes are flying horizontally towards us and they turn together, greeting us in the famous Diamond Bend formation, which usually starts the Red Arrows show. Fabulous.

Now three of the planes have split off and ... ah yes, two of them are climbing. Now, they cross at the top, bank and dive towards each other to make a Heart shape with their smoke ... Oh, and here comes the third aircraft right through the middle to complete the picture ... fantastic!

The three are joined by a third aircraft now as they fly away from us. I wonder what - ahaa I think I can see what - yes, now this really is amazing, ladies and gentlemen. Watch closely. Two of them are rolling in long horizontal loops in a kind of spiral round each other, just like a corkscrew, while the other two fly straight and flat inside. They're so close to each other, each pilot is only three or four metres away from the pilot next to him during this manoeuvre, and there's no computerized flying here let me tell you - every movement of the control surfaces is under the direct manual control of the pilot. That manoeuvre is called The Corkscrew in fact. Very demanding.

The Red Arrows, of course, spend the whole of the winter each year practising for the display season – six or seven months in all. Every pilot always flies in the same position in the formation. That group has headed off and been replaced by another group of four and they're climbing ... climbing ... climbing. Now, they're looping over, and down they come, diving down vertically ... that's extraordinary - I really don't know how to describe the shape they've drawn ... it's like straight vertical lines with loops at the bottom. The team have given it the equally strange name 'Twizzle'. I bet that one takes a lot of practice!

What's next, I wonder. Here we've a bigger group, flying horizontally straight at us again, two, four, six, seven planes flying close together wooah and they suddenly break apart. What a beautiful sight, all the coloured smoke trails rushing away from each other. The pilots experience 7 g of gravitational force ... ladies and gentlemen in that turn – the Vixen Break. The Hawk aircraft itself has a structural limit of 8 g!

And now a smaller group is climbing for the Opposition Loop. Up, up. And they turn together in very tight formation ... accelerating downwards. Down they go. Beautiful. And they spread apart just a little but still stay close to each other. These pilots really are exceptional.

Unit 5

Unit 4, Lesson 2, Track 19

R: Now you've flown all kinds of planes, haven't you, Tom?

T: Yes, it's nearly 30 years since I started and I hope I'll be doing it for another 20!

R: What changes have you seen?

T: Well, I suppose the biggest change really is in engines. Aircraft body design hasn't altered so much, but engine technology's another matter. And it's the power source that drives the machine along, so it's central to the whole thing. Without that, it's just a glider.

R: That's true! What would you say are the important factors in engine design?

T: I'd say probably the most important thing, above all others, is reliability. You have to remember that a plane can't just stop in the air if something goes wrong, like a ship or a car can do. You need to be absolutely sure that the engine will keep working from the time the plane takes off to the time it lands. The thing is that aircraft engines have to run at very high power most of the time. When it takes off, of course, it's using maximum power, but even when it's just cruising along happily it may well need between 65 and 75% of its power - whereas a car only uses that percentage of power for about 20% of the journey time.

The next most important factor in my book is weight: the lighter you can make the engine, the better. If the plane is carrying less weight, less power is needed and less fuel is required. So reliability and weight, yes, those are my top two. In the early days, engines tended to be pretty heavy, but these days, with lightweight metals like aluminium, and other technical improvements, they are much, much lighter. They've got some incredible materials these days.

R: What else would be on your list apart from those two?

T: Well, the next priority I suppose would have to be power - the more power in an engine the better, because then the weight and the air resistance can be overcome easily. Plus, it's always good to have some power in reserve in case of an emergency situation. And lastly, I guess, well size makes a difference. The smaller it is, the less drag you get again, the less air resistance it has, and if you can streamline the shape, that helps too.

R: Sure. And wh- ...

T: Actually, there is one more thing, sorry, if it can possibly be done. It's nice if the engine is kept as simple as possible, with spare parts that are easy to get hold of and reasonably cheap. There's nothing worse than having your plane grounded because some high priced part has to be sent from the other side of the world.

Unit 4, Lesson 2, Track 20

R: Which engine is your favourite?

T: My favourite … well, there are three that I've had good experience of. The twin cylinder in-line, you know, one behind the other, is a nice engine. It's a fairly simple design and, of course, a very well streamlined arrangement, and as I said before, that's always a good thing, although it doesn't have much power compared to some other types. Then there's the radial with the cylinders set in a circle. That's a bit more complicated. More spare parts of course, but it is very powerful and it runs very smoothly. And I guess the other would have to be the gas turbine. A turbine is always expensive but it's powerful and again, it's a very good shape for fitting in a plane. No cylinders, of course - the whole thing is very streamlined, so drag is kept to a minimum.

R: Which was your first engine?

T: Oh none of those! No, I started like most people of my age with a rubber band! I didn't have the money to buy a proper engine. No, I just turned the propeller with my finger a couple of hundred times and then let the plane go! Ah … happy days.

Unit 4, Lesson 6, Track 21

I've always been interested in aircraft and in engines particularly, so when I retired, I decided to see if I could find an old aircraft engine to work on. Well, I advertised in local newspapers as well as a couple of aircraft magazines, and nothing happened for a month or so. And then out of the blue, when I thought I wasn't going to get anywhere, I got a phone call from a film company. Apparently, they had an old Rolls Royce Merlin engine in working condition. They were using it to produce wind in the studio - winds of up to 400 miles per hour, they said. But it was unreliable as well as being noisy and smelly, so they were going to replace it with more modern equipment. They didn't want to just throw the engine away, so I could have it free if I paid for the transport. I didn't hesitate. And a week later, I was the proud owner of a working - well sort of working - Merlin MK 20 aero engine.

But they weren't joking when they said it was unreliable. In fact, it was really on its last legs and should have had a complete overhaul. Anyway, the first time I started it up, steam came hissing out at the front, the two front cylinder heads, because of a water leak. Luckily it was quite easy to fix. I just fitted new cylinder head gaskets. The next problem was another leak. This time oil, not water. There are these brass tubes that carry oil back to the crankcase and they were all leaking … not much, but enough to be a problem. They all had to be replaced … And that was quite expensive, because I had to have them specially made. Spare parts for 50-year-old engines aren't that easy to get hold of.

Things went all right for a while … until I started to get problems with the carburettor. It was flooding, you know, filling up with too much fuel, so of course the engine wasn't running properly, because the fuel was staying in the carburettor, you see, instead of going to the combustion chambers. Anyway, that's what I thought. Well, I stripped down the carburettor, inspected it, it looked OK, but I cleaned it anyway, and then put it all back together again, started the engine, and it was just the same! It still wasn't firing on all cylinders. So I removed the rocker cover and what did I find? One valve completely missing and half the valve springs broken. So I replaced all of those, including the unbroken ones, of course. After that, I didn't have any trouble for about a year, and then the two front cylinder heads started leaking again. This time I completely dismantled the engine. And it was then that I discovered that it had been in some sort of crash and the front cylinder blocks were slightly out of alignment. Anyway, they had to be machined so that they fitted properly. While I had the engine in pieces, I completely overhauled and replaced anything I could. Since then, she's been running perfectly. Listen …

Unit 4, Lesson 8, Track 22

It is easy to forget what a technological marvel modern flight is. When a Boeing 747-400 is cruising at 35,000 feet, each of the four engines is generating 12,000 pounds of thrust, and continues to do so for many hours at a stretch.

Take-off **Tapescript**

To do this, the engine draws in 700 pounds of air every second, although, in fact as much as about 80 percent of the air which goes in actually bypasses the core. The 120 pounds of air that does enter the core is pressurized to more than 150 pounds per square inch and at the same time, heated to more than 850°F in the compression section.

When the air is compressed inside the engine core, one-and-three-quarter pounds of fuel is injected into it and burned to heat the air/fuel mixture of combustion gas to more than 2,000°F. The turbines extract enough energy from these gases to turn the fan at about 3,300 rpm and the compressor much faster still, at around 9,500 rpm. When the gas mixture exits the turbine section, it is moving at a velocity of 1,400 feet per second and is still very hot - over 1,000°.

All this goes on every second that the engine is cruising up there, hour after hour.

Unit 4, Lesson 10, Track 23

I: Ok now we talked about fuel and fuel tanks in small aircraft yesterday. Can anyone remind us what the two main problems are with aircraft fuel? Yes, Ali.

A: It's highly flammable, so there is always a danger of maybe fire.

I: Yes, that's right. But of course, the trouble is if it wasn't highly flammable, it wouldn't be much use ... But you're right, you have to design everything to get round that problem. What was the other main problem - there's something else about fuel which can cause problems if it isn't allowed for. Yes, Abdelhakim?

A: It's a liquid, so it's a bit difficult to store. It can run around the plane.

I: Yep, that's it. Not exactly around the whole plane, of course, but round –

A: In the tanks.

I: - yes, you've got it. That was it. The fuel can move around inside the fuel tanks when the plane turns - you just try running round the canteen with a bowl of soup in your hands. It'll

go all over the place, especially if you suddenly stop, start or turn. OK fine, so how do you think this causes a problem for a pilot flying a plane?

A: It makes the plane difficult to keep balanced because the weight of the fuel keeps moving in position. It's like all the passengers suddenly change seats.

I: Exactly. And the plane becomes very difficult for the pilot to control. So what do you think the answer is?

A: You have to keep the fuel balanced all the time you are flying.

I: That's right. And how can you do that? Ali?

A: A lot of smaller fuel tanks?

I: Well OK, yes. Quite hard to build though. Other ideas? Mohammed.

M: Move the fuel from one tank to the other while flying.

I: That's right. Well done. How?

M: A pump?

I: By pumping it from one tank to another ... in any part of the plane, in any of the tanks, and so you maintain the balance of the plane – the 'trim', as it's called. For example, you don't want a right wing tank empty and the left one full. And of course the bigger the aircraft, the more important this is ...

Unit 4, Lesson 10, Track 24

I: ... and what kind of pump, can anyone suggest ... I mean, there are different kinds, so which is best for moving fuel in and out of tanks? Yes, Yusef.

Y: It has to be very safe.

I: That's true. Well ... Have a look at these pictures. Now these five are pumps from a trainer plane, and they all do different jobs. Let's look at the first one - pump number 1. Can anyone tell me what type of pump this is?

Y: A gear pump.

I: Well done, Yusef. In fact, it's the high-pressure pump, which supplies the combustion chamber of the engine. You can see it's got a filter on the outlet to make sure that the fuel is clean when it goes into the engine. Yes, Abdelhakim?

A: Where does the power come from?

I: From the engine. It's connected to the engine. As the engine turns, the pump rotates and supplies more fuel.

Y: But how it can start if the engine isn't turning, before the plane takes off?

I: Good question. And the answer is this second kind of pump. This doesn't need any engine power. It's a 28 volt booster pump which runs from the battery until the engine takes over and powers the high pressure pump. And it's a centrifugal pump. In the PC9, there is one in each collector tank. Those are the two fuel tanks in the centre of the plane. You can see from the flange at the bottom of this pump that it's fixed to the bottom of the tank. Yes, Ali?

A: Isn't it dangerous to have an electric pump in a jet fuel tank?

I: Well, no, not really. You see, it is the fuel vapour that is most dangerous, and these booster pumps are immersed in the liquid fuel. Also, they are carefully sealed and they are not used for long periods. Now, the pump that is used for long periods is this third one here, the main fuel pump. You could call it the 'heart' of the fuel system.

Y: It looks like a vane pump.

I: Yes, that's right, that's exactly what it is. It's got rotating vanes in the centre here, which push the fuel through. It's powered by the engine, like the gear pump.

A: What's the valve for?

I: The one on the right you mean, Abdelhakim?

A: Yes.

I: Ahh, that's to relieve the pressure if it is too high on the outlet side.

A: I see …

I: This vane pump does two jobs. It not only supplies fuel to the first pump, the high pressure gear pump. It also drives these other two pumps, numbers 4 and 5. As you can see these two are quite similar. Let's look at the simplest one first, number 4. What do you notice about it?

A: There's nothing in it, just a tube.

I: Right, and pump number 5 is the same, except

for this flap valve at one end to make sure that the fuel can only go in one direction. They are both called jet pumps and they work on the 'venturi' principle. Now then, can anyone remind us what that is? Yes, Yusef.

Y: When a high speed liquid goes through a smaller space it makes a vacuum behind it.

I: That's it. You see the main pump, number 3 here, drives the fuel, in what's called the motive flow, through the main tube of the jet pumps, and then more fuel is sucked up by vacuum. This type, number 4, is used to transfer fuel from the wing tanks to the collector tanks. It's called a transfer jet pump. And number 5, this one with the valve, called the delivery jet pump, supplies fuel continuously from the collector tank.

A: And they work all the time as the plane is flying?

I: They do. No moving parts, not much to go wrong and they can keep working indefinitely, non-stop. So the fuel is always moving from the wings to the centre and keeping the plane properly balanced. Yes, Abdelhakim?

A: Can you show us one of each type of pump?

I: Yes. In fact, we're going to strip a couple down in the workshop tomorrow.

Unit 6

Unit 6, Lesson 4, Track 25

Part 1

Mr G: Hello, John. How are you getting on with drilling out those two sheets of aluminium?

J: Fine, thanks, Mr Green. We've drilled out most of the holes. I'm doing the outside holes along the edge on the machine here, but we can't do the centre holes on the machine. Frank's doing those with a hand drill over there on the bench behind the guillotine.

J: Frank, Frank … are you ok? You're as white as a sheet. What have you done to your hand?

F: No, I'm not ok … I think I've broken my little finger …. The chu-… aaghhh the chuck was still in the drill … ooohhh … after I changed the bit, … I didn't notice …

x

Mr G: You've cut it quite badly as well. Right, let's get the first aid kit and get the finger covered and then off to hospital ASAP. You're going to be off work for a couple of days, even if it isn't broken. I'd like you to go with John. Report back to me when you get back, ok?

J: Right. Shall I ...?

F: So stupid ... aaaa ... I just didn't check ... I looked away for a second.

Mr G: Ok, Ok, we can talk about it later, but now let's get that finger treated. That's the first thing.

Part 2

Mr G: Come in.

Ah, hello, John. How did it go at the hospital?

J: He was quite right, the finger is broken, but it's a nice clean break. The doctor said he ll need six weeks off work though, because he works with his hands. It would be different if he was working in an office.

Mr G: Right ... well, we'd better do this form. Just check it with me. So, reference number. Workshop 2, wasn't it?

J: Yes.

Mr G: So, W2. 9 - dear me, the ninth accident this year – 06. OK, ... full name. Could you just read off his personnel data up on the screen there?

J: Yeah, Frank ... no sorry, his full name's Francis f-r-a-n-c-I-s. Yes, Francis Robert - I never knew that. Robert ... – the surname's Day ... address: 245, Bartlett Street.

Mr G: Is that double t at the end?

J: Yes. Bart b-a-r-t and lett l-e-double t.

Mr G: Occupation ... fitter. Age?

J: 28.

Mr G: And he works directly for the company. Type of injury ... fracture. First aid ... 'yes'.

J: Yeh, we put a loose bandage round the whole hand.

Mr G: OK ... agent of injury ... that's clear ... and injury site ... left finger. Date and time ... 26th November, and it was, what ... about 11 o'clock?

J: 11:15.

Mr G: Yes, that's right, and you and I were the witnesses, so I'll put my name and address. Now, cause of incident. Hmmmm ... I don't think we can put that in until I've spoken to Frank a bit more. Just to make sure we get the facts straight. Now, brief description of accident. I'm going to put 'operative injured by chuck key left in hand drill'.

J: Yeah, that seems to be what happened.

Mr G: Hmm ... recommendations ... No, again, I'll wait until Frank is back before I do that. Now, date of incapacitation ... same as today's. Date of return ... can't say yet. I think that's all we can do for the moment. Oh yes, property damage.

J: Well, the alclad sheet he was working on can't be used again.

Mr G: No, you're right ... I'll talk to the accounts department about the cost of that. Right, that's it for now. If you can just put your name and address in the top right hand box and I'll send this across to the safety superintendent ...

Unit 6, Lesson 8, Track 26

Mr A: Hello, 201 70973.

Ted: Is that Mr Armstrong?

Mr A: Yes, speaking.

Ted: Oh, hello sir. It's Ted from Sankey's garage here. Just ringing to let you know your car is ready for collection. We've done the full 20,000-mile service as you asked.

Mr A: Everything ok?

Ted: Yes, all pretty good, really. I'm just looking at the job sheet now ... Ah, yes ... except for - you said the brakes didn't feel right ... Yes, we had to do quite a lot on those as it turned out.

Mr A: Oh, so that's going to be quite expensive.

Ted: Well, I'm afraid it won't be cheap. But the rest of the service was fine.

Mr A: So, what was the problem?

Ted: There were a couple of things, actually. First of all, there was a leak in the master cylinder, brake fluid was getting out and air was getting in.

Mr A: So, did you repair it?

Ted: Oh no, now it would be expensive to repair the old one. No, it's cheaper to replace them. So we did that ... we also found a slight leak in the fluid line to the left hand front brake. So, obviously we had to empty all the fluid out of the system to replace the damaged tube.

Mr A: And you refilled it with new fluid.

Ted: Yes, and I'm afraid that we also found the surface of the disc on the right front wheel was damaged - it was slightly out of alignment.

Mr A: Why was that?

Ted: Hard to tell. It's usually caused by excessive heat, you know, if the brakes are having to do too much work ... When we inspected it, it looked as if - have you been doing any hard or fast driving recently?

Mr A: Yes, we spent the week before last driving in the mountains. That's when I noticed there was the vibration and the car was pulling to one side when I pressed the brake pedal. You have to do a lot of braking on those roads.

Ted: Sure. Yes, I should think that front wheel brake unit was doing most of the work because the other brake wasn't getting sufficient pressure. Anyway it's all fine now. We machined the damaged disc surface and straightened it again. We've checked and cleaned the other three discs and they're all OK, so it's ready for you to collect and working perfectly.

Mr A: Mm ... that's the good news. What's the bad news?

Ted: Ha ha ... well, yes, the main service was the standard £120 plus parts, but then the extra work on the brakes is going to ...

Unit 7

Unit 7, Lesson 2, Track 27

1.

A: Do you know where Jack is? I've been waiting for him for twenty minutes.

B: Last time I saw him, he was going off to wash his hands. He's been doing some painting ... Oh, here he is now. All right Jack? Why the bandage?

J: I've scalded myself with that water. I should have put some cold in first. It serves me right for ignoring the notice! Lucky it's only one hand.

A: You won't make that mistake again in a hurry.

2.

Foreman: How's that pump running now, Barry?

Barry: Fine, I've replaced the front main bearing so I'm just going to give it a bit more lubrication, clean it up and then it's ready to re-fit.

Foreman: Ok ... but make sure it's switched off when you do it! We don't want another accident like last week.

3.

A: Phyeeugh ... look at it, this overall's ruined and I've got some down my neck, I ... can feel it. Ugh, its all sticky.

B: And some on your face and hair, too ... look at the state of you ... what happened?

A: The sump plug fell out while I was underneath ... It's not ... funny!

C: It is from where I'm standing!

B: Well, you're completely covered in the stuff ... off you go to the washroom and clean yourself up thoroughly ... and make sure you change that overall!

4.

A: Ah ... Yes, I think this is the right stuff for cleaning all that paint away.

B: Better check first, though. What does it say on the label?

A: Hmmm ... harmful, do not swallow. Only use in a well ventilated area. Wash hands immediately after use. No smoking when using this product. Do not breathe fumes. Keep in a cool place well away from sources of ignition ... Here we are ... To use as a paint solvent, apply to a cotton cloth.

5.

A: Have you finished the block yet?

B: Yes, I'm just going to finish it off on the grinder ... get it nice and shiny and smooth.

A: Ok, but make sure you don't try to use the grinder over there. There's no protection on it at all. It got broken yesterday afternoon and hasn't been replaced yet. In fact, I'll go over and remove the wheel so it can't be used.

B: Might be an idea to disconnect it from the power supply as well.

A: Yep ... I'll do that.

6.

A: So this is where the new milling machines are going to go.

B: Yes, they're being installed next week ... once this place has dried out.

A: Hmmm ... yeah it's a pretty strong smell. They could do with a few more windows open. So what's going to go in these cupboards, do you know? Look out, mind that door.

B: Oh no! This is a new jacket. There should be a sign here.

Unit 7, Lesson 4, Track 28

A: Hello and welcome to *I could do that,* the programme in which we give young people a chance to ask questions about interesting and unusual jobs and today in the studio my guests are Martin Robbins. You're how old Martin?

M: 12.

A: ... and Terry Gardner from the Civil Aviation Authority. I won't ask you how old you are Terry.

T: Thank you.

A: Now you are, I think I've got the title right, a civil aviation operations cabin inspector.

T: That's correct.

A: And can you give us an idea of what you do, in a few words?

T: Well as the job title suggests, I inspect aircraft cabins and make sure that the correct procedures are followed by the crew before, during and after a flight.

A: Right. Now I think it's over to you Martin. Let's have your questions.

M: Can you tell me - what exactly do you actually inspect?

T: Most of the things that you would expect to find inside an aircraft really -

M: Like the engines and the cockpit and stuff?

T: No, no, I'm not an aircraft engineer. Those sorts of thing are checked by the technical maintenance people. No my main priority, the most important thing I mean, is the safety of the aircraft passengers - making sure that everything in the passenger cabin is OK and that the cabin attendants do all the right things throughout the flight. For example one of the first things I check when I go on board is that there are legible signs in all the correct places. You know the sort of thing - No Smoking, Exit signs, emergency equipment. To make sure the signs are there and that everyone can read them easily, especially in case of an emergency.

M: What sort of emergency equipment do you look for?

T: Well, every model of plane is different of course, there's a different specification for each one. But I need to make sure that there are the right number of fire extinguishers, that they are in the right place, and that they have an up-to-date service tag.

M: What's that?

T: The service tag. That's a little label which tells you when the equipment was last checked. And of course there's a lot of other safety equipment. Aircraft – or airlines - are not allowed to fly if they don't have all the right safety equipment. Things such as oxygen bottles, and oxygen masks - protective breathing equipment in case there is smoke in the cabin. It's important to check that they that drop correctly down from the ceiling. Then there's First Aid kits ... and the passenger information cards, of course ... the flashlights - the torches – I need to make sure there are enough of them – in case the lights go out and, of course, I check that the batteries are charged.

M: Are the information cards the ones in the back of the seat with the pictures on, with the magazine and the sick bags?

T: Yes, that's it, and those are important, those sick bags, especially if there's a lot of turbulence - you know, if it gets bumpy - for everyone's comfort, not only the person who feels sick, and to keep the cabin clean.

M: Yuck. Is that the lot?

T: No! There are dozens of other things … Can you think of anything I might have missed?

M: What about lifejackets?

T: Well done. I have to check that there's one under each seat. And again I check the condition … Oh, yes, of course there has to be at least one megaphone.

M: In case the intercom on the plane doesn't work.

T: Yes, for that, but mainly for use if the plane has to be evacuated and the crew need to give instructions to the passengers.

M: So after you've checked everything is there, the plane takes off?

T: Not quite. I inspect the cabin before the passengers come on board. And before they do, there's another very important thing I have to check. I can't really do it with the passengers on board - any ideas?

M: Uhh … mmmmmm … oh, I know - the seats.

T: Exactly. I check that they're working, that they can be adjusted up and down, and that the seat belts work properly and that they're not twisted or frayed. I also check the cabin crew seats and their safety harnesses as well. There really is a lot to look at. Then, when I'm satisfied that everything's ship-shape, the passengers are allowed to board.

M: Do you sometimes go on a flight without the crew knowing that you are there?

T: Well … actually, I'm afraid that's confidential. It's the one thing I'm not allowed to tell you.

A: Ha ha. I think we can take that as a yes. OK, now perhaps you could tell us a bit more about the next stages in the inspection.

T: Well, as the passengers come on board, I make sure that the cabin crew show them to …

Unit 7, Lesson 8, Track 29

During my last two years in the US Navy, I served on board an aircraft carrier as a Quality Assurance Representative. For those of you who don't know, the job of the QAR is to keep an eye on the standard of the maintenance, servicing and repair of aircraft.

When I first joined the ship, I was pleased to discover that I wasn't going to be dealing with LOX converters. I don't know if you know, but these are a specialised kind of cylinder assembly designed to store and convert liquid oxygen into gaseous oxygen for aircrews during flight. Hence the name LOX: liquid oxygen. Well, there were a lot of maintenance problems associated with that system and I was glad that the Hornet fighters based on the ship were all in fact fitted with the new OBOGS oxygen system. OBOGS stands for On Board Oxygen Generating System, and it was known to be much more reliable. It works by concentrating oxygen from the air during flight and supplying it to the pilot at high altitude: there is no need for the aircraft to carry the oxygen supply.

Unfortunately, a few weeks after I joined the ship, some of the pilots began to report experiencing symptoms of altitude sickness - such as nausea - on a couple of the Hornets, and the OBOGS system was obviously the first thing that was looked at. I consulted all the relevant maintenance manuals carefully, read about similar incidents, and of course questioned the pilots who'd reported the problem. One of the ship's technicians inspected the OBOGS systems and found that on deck it was working perfectly. So, knowing the system worked fine on the ship's deck, but not at altitude, we decided to test the concentrator to see if it deteriorated at high altitude. But no, it was fine. Absolutely no defects.

However, some time after this, one of the aircraft mechanics came to see me and showed me an OBOGS line he had removed. A closer look revealed the hose's exterior metal braided sleeve had separated from the plastic tube inside, which meant that we could see the true condition of the internal hose. We could see that it was almost completely closed off, blocking the free flow of oxygen. The damage was hidden from casual examination because of the protective metal sleeve on the hose. We quickly recognized this was likely to be the source of the problem, so I then went to maintenance control and asked to inspect all the other aircraft. We wanted to determine if this was an isolated case or a general problem. After inspecting 12 aircraft, I found nine OBOGS lines bent or crimped in some way, possibly causing low oxygen.

We now tried to find out the reason for the crimping. I followed our mechanics out to an aircraft and watched them remove an OBOGS concentrator. I quickly discovered that their technique didn't appear correct. The service manual said to move the product line to the side, which allowed space to remove the concentrator. However, this technique resulted in the mechanics treating the OBOGS lines like the old LOX lines: they just moved them out of the way quite carelessly. The difference was that the LOX lines, which we could move aside quite easily, were very flexible, whereas the new OBOGS inner lines were made of a much more rigid plastic. So when the mechanics moved the OBOGS lines quite forcefully, they were getting crimped - and they were crushing the inner plastic hose. The mechanics thought that the line was as tough as the LOX line and couldn't be damaged.

The result of the crimping was that there was some back pressure within the obogs concentrator. So it formed a blockage - which reduced the amount of oxygen going to the pilot. The reduced amount of oxygen didn't affect a pilot's performance until the aircraft reached very high altitudes, which is why the system seemed fine on the ground. A final piece of this puzzle could have resulted in death! In those early days, the OBOGS oxygen monitor only detected the oxygen concentration within the concentrator, which meant that a warning light would not go on because the level inside the concentrator was normal. An oxygen monitor didn't exist downstream, so the pilot never got a warning of the oxygen decrease.

Well, what we learned from this was to stop treating OBOGS lines like LOX lines and to be much more careful with them when they were moved. And I'm sure we saved the lives of a few pilots as a result.

Unit 8

Unit 8, Lesson 1, Track 30

A: Right, let's have a look at this then.

B: So why did they request a tyre inspection? We don't normally do it every time the plane lands, do we? Usually one of the crew does a quick visual.

A: No, I know, but apparently she had a bad landing because of these sudden crosswinds. Came down with quite a bump. And skidded a bit more than usual. That can sometimes do a lot of damage when the plane comes down quite hard.

B: Ok, so what's the main thing to look for?

A: We'll do it by the book. Have you got the tread gauge there?

B: Here you are.

A: Thanks. Hmmm … Right, let's see. Just put it in the grooves - there … and there … and there … Yep, that's good. There's plenty of wear left on these tread. So let's just stand back and look at it from the front. Ah yes, small problem there, can you see?

B: Oh yes, the treads look a bit more worn on the right hand side than on the left. Does that mean replacing it?

A: Not necessarily. If there's no other problem, we can just flip it.

B: Flip it?

A: Demount it and turn it round so that the wear is evened out. But we'll need to report it, so that the nose wheel gear is checked.

B: So that might be the cause of the problem. Gear misalignment.

A: Yes, and it's probably causing a bit of nose wheel vibration as well. Right, now we need to look for anything stuck in the tyre. No … Looks OK. Can you see anything? Any foreign matter?

B: No. Looks all clear. What happens if you find something stuck in the tyre? Do we try to remove it?

A: Not while the tyre is inflated, you don't. You could burst it and really hurt yourself. No, if there's anything noticeable, it has to come off and be repaired in the workshop. Now, let's have a look at the beads and the sidewalls.

B: What are we looking for?

A: The obvious things really - heat damage on the beads, they're the hottest parts of the tyre. Cracking or bulges.

B: Like this you mean?

A: Ah yes. Well spotted. It's only about the size of your thumbnail, but it's a definite bulge all right. Well that's it then, off it comes. The shop'll have to inspect it, no question.

B: But it doesn't look too bad.

A: Not yet, it doesn't, but it's a bad sign. It means that there's some damage or separation between the layers. Every time it's used, it'll get worse. Right I'll let the flight crew know what's happening and then we'll get started.

Unit 8, Lesson 3, Track 31

F: Good morning,

B: Mr Sturgis?

F: Yes, that's right. We've got a flight booked for today – it's my son's birthday.

B: Hello, James. Happy birthday. I'm Bob.

J: Hello.

F: I've been promising him a balloon flight for ages – a couple went over our house a few months ago and he hasn't stopped talking about it since.

B: Yes, they're a pretty impressive sight. It's the size of course - and the colourful designs.

F: How big are they actually?

B: Well, the one we're going in today is 10 meters across and about 20 meters high, complete with the basket and gas burner.

J: It doesn't look very big.

B: Not yet, but you wait till it starts to fill. Ah, there's my co-pilot Simon. He's just going to start filling it. He'll start with a petrol engined fan to start off and then when the mouth of the balloon is open a bit wider, he'll light the propane burner.

J: Look Dad, that blue one's starting to go up. Wow, it's enormous!

B: Now, while Simon's inflating the envelope, the main balloon, are there any questions you'd like to ask?

F: How does it actually work?

B: Well I don't know if you remember something from school science called Charles' Law.

F: Hmmm … vaguely. Science wasn't really my strong subject at school.

B: Right, there goes the burner. Shouldn't be too long now. Yes, Charles was an 18th century French Scientist, who made the discovery that the volume of a gas is directly proportional to its temperature, so his law states quite simply that if you increase the temperature of a gas - which is what young Simon is doing over there - then the volume will increase, as you can see starting to happen.

J: Oh yes, look Dad, it's starting to fill and move up. It's nearly upright.

B: That's because a certain amount of hot air occupies a larger volume than the same quantity of cold air. And since it occupies a larger volume, it must be less dense than the same amount of cooler air. And less dense means lighter. Yeah? Does that make sense?

F: Um.

B: So, when the gas in a balloon gets hot enough, the weight of the balloon with this hot air is less than the weight of an equivalent volume of cold air, and the balloon starts to rise. Yes, another ten minutes or so and we should be able to get on board. The four of us'll fit in the basket.

B: Right. Everybody ready?

F: Yes.

J: Yeaaah.

B: Good. We just release the ropes that hold the basket down, and - off we go.

F: Woah. It's so strange. The ground is getting smaller, but it doesn't feel like we're really moving.

J: All the people are getting smaller.

J: This is fantastic. Dad have we got to come down yet?

F: I'm afraid so. We only booked a half-hour trip.

J: How do you make it go down again?

B: Well, of course, I stop using the burners. That way, the air in the balloon cools down, so it contracts and becomes more dense – heavier, if you like, and so we start to sink. Exactly the reverse of the process that made us rise.

F: How long does that take?

B: Well, it depends on the difference between the temperature inside and outside the balloon. The bigger the difference, the longer it takes. But I can speed things up a bit if necessary if it's taking too long. If I pull this cord, you can see that flap at the top of the balloon opens – and it lets some of the warm air out, so the volume and the temperature are both reduced suddenly, so we'll start descending more quickly.

J: I can hear the birds again.

B: And here we are.

J: Fantastic. Can we come again another time? Please Dad.

F: Well, I'll have to think about that. It's not cheap you know.

Unit 8, Lesson 5, Track 32

Ali: Excuse me.

Mr B: Yes Ali.

Ali: I'm looking for information about heat pumps. Could you tell me something about how they work?

Mr. B: Of course. How much do you already know?

Ali: Well, I know that a heat pump transfers heat from one place to another … but the process isn't very clear to me.

Mr. B: Well, OK, basically a heat pump uses the fact that when a liquid evaporates, it needs to absorb heat from its surroundings and when a vapour condenses, it gives off heat. The best example is an ordinary domestic refrigerator. The inside is cold, but the grill at the back is warm. The heat is removed from inside the refrigerator and then given off into the room. Which is why you can't cool down a room by leaving the refrigerator door open.

Ali: Because when you leave the door open, the refrigerator works hard to try to cool the room, and at the same time, it's giving off the same heat back into the room, heating the place up.

Mr. B: That's right. The heat can't just disappear. It's transferred from one place to the other – and the room doesn't get any cooler. You just waste a lot of electrical energy.

Ali: But I'm not sure what the main parts are, or how they are joined together.

Mr. B: OK, well, in a refrigerator, or any heat pump, you need a compressor, an evaporator, a condenser, a control valve and a special fluid called a refrigerant. Look, there's a drawing in the book here.

Ali: The evaporator must be the part where the refrigerant changes from a liquid to a vapour.

Mr. B: Yes … here it is - and remember that it needs to absorb heat to change from the one state to the other, from liquid to gas, and it absorbs the necessary heat from inside the refrigerator, and the inside of the refrigerator gets colder.

Ali: So the condenser is where the refrigerant changes back from vapour to a liquid, and gives off the heat again.

Mr. B: Yes, that's right. It sort of gives the heat back - in a different place - at the back of the fridge. The heat from inside the fridge ends up in the room.

Ali: OK. But what about these valves on the pump? Are these valves - V1 and V2?

Mr. B: Yes. These valves operate alternately. When the piston in the pump moves inwards, on the compression stroke, valve two opens and valve one closes.

Ali: Is it correct that when the piston moves inwards, the vapour is compressed into a liquid and forced through valve two into the condenser?

Mr. B: Yes, it is, and on the induction stroke, when the piston moves outwards you get the opposite - valve one opens, and valve two closes. So the pump keeps the refrigerant circulating by opening and closing these valves. Of course, the refrigerant has to be a special fluid that can change from liquid to vapour and back again quickly and easily.

Ali: And what does the control valve do?

Mr. B: OK, think about the pressures in the evaporator and the condenser. Which will be higher do you think?

Ali: Ah, the pressure will be higher in the condenser than in the evaporator, because it's gas in there.

Mr B: Exactly. And it's the control valve, sometimes called an expansion valve, which maintains the difference. Basically, it's a small orifice, a small hole, which forces the liquid to change into a vapour in order to flow through.

Ali: So it's too small for a liquid to go through, but big enough to allow a vapour through?

Mr B: That's it.

Ali: Hmm.

Mr.B: Is it all a bit clearer now?

Ali: Yes … Yes. Thank you very much.

Mr B: You're welcome.

Unit 8, Lesson 5, Track 33

Well OK/ basically a heat pump uses the fact that when a liquid evaporates,/ it needs to absorb heat from its surroundings/ and when a vapour condenses,/ it gives off heat./ The best example is an ordinary domestic refrigerator./ The inside is cold,/ but the grill at the back is warm./ The heat is removed from inside the refrigerator/ and then given off into the room,/ which is why you can't cool down a room by leaving the refrigerator door open./ Because when you leave the door open,/ the refrigerator works hard to try to cool the room/ and at the same time it's giving off the same heat back into the room,/ heating the place up.

Unit 8, Lesson 9, Track 34

OK, I think we're ready to start. OK, what I'll do is to give you an overview of the system and then go back into it in more detail, with some numbers and statistics.

OK, so, pressurized air for the cabin comes from the compressor in the aircraft's jet engines. Moving through the compressor, the outside air gets very hot as it becomes pressurized. The part drawn off for the passenger cabin is first cooled by heat exchangers in the engine struts and then flows through ducting in the wing. After that, it's further cooled by the main air conditioning units under the cabin floor.

The cooled air then flows to a chamber where it is mixed with an approximately equal amount of highly filtered air from the passenger cabin. The combined outside and filtered air is taken to the cabin and distributed through overhead outlets.

Inside the cabin, the air flows in a circular pattern and exits through floor grilles on either side of the cabin or, on some airplanes, through overhead intakes. The exiting air goes below the cabin floor into the lower lobe of the fuselage. The airflow is continuous and quickly dilutes odours while also maintaining a comfortable cabin temperature.

About half of the air exiting the cabin is immediately exhausted from the back of the airplane through an outflow valve in the lower lobe, which also controls the cabin pressure. The other half is drawn by fans through special filters under the cabin floor, and then is mixed with the outside air coming in from the engine compressors. These high efficiency filters are similar to the filters used to keep the air clean in hospitals. They are very effective at trapping microscopic particles as small as bacteria and viruses. It is estimated that between 94 and 99.9 percent of the airborne microbes reaching these filters are captured.

OK, now as I said, some more detailed numbers …

Unit 9

Unit 9, Lesson 3, Track 35

T: Mr Martin?

M: Hello. Yes.

T: How do you do. I'm Gerry Townsend. I'm the inspecting technician for the electrical system in your plane. I've just been having a look.

M: Oh right how was it? Everything in order, I should think.

T: Well, yes and no. 95% is OK, but I noticed that you've recently had some new equipment installed by another company.

M: Yes, that's right. Last month, I had a new radio unit fitted. Works very well so they must have done a good job, surely?

T: Oh everything is connected up the right way. But

to be honest with you, I'm not really happy about the standard of the wiring. There are a few things that could be a safety problem.

M: Really.

T: Yes, I'm afraid so. I can go through it with you here, but it'd be better if we went out to the plane so I can point things out to you.

M: OK, fair enough, let's go. I'll just get my coat. It's a bit nippy out there.

T: OK, right here we are then. Let's go point by point. Now, they have used the right cable, no problem there. You can see from the identification number marked along the outside sheath ... this number tells me the size ... the cross-sectional area of the conductor, and this number at the end tells me it's stranded copper with vinyl insulation, and that's correct for equipment with this sort of power load.

M: Gooooood, so far.

T: And I'm quite happy about the route of the cables. They're tucked out of the way, so nobody's likely to walk into them or hang on to them by mistake. They've also remembered to fit them above the hydraulic fluid line, just in case there's a leak. So that's all fine.

M: OK, so what is the problem then?

T: Well, if you look at the way the connection has been made to the terminal block -

M: Looks very neat and tidy.

T: - yes and that's actually the problem. They haven't left a loop. There should be an extra six-inch diameter loop of wire to allow for any sudden tension in the cables. It also makes everything much easier to take out and put back if there's extra wire available.

M: Oh dear.

T: These are also the wrong type of lugs to connect to the terminal block. They should be the closed ring type, not these open-ended ones – these can slip out too easily if there's a lot of vibration.

M: Yeah, that makes sense.

T: Now the wires have been passed through this hole at the side and that's fine - the hole is plenty big enough. But there's no grommet to stop chafing.

M: And what about where it comes through the other side? That clamp looks a bit too tight to me.

T: Yes, you're right. In fact, it's actually pinching one of the cables. See, it could eventually cut through the sheath and into the insulation sleeve, and look here where the cables are hanging down between the clamps. There's far too much slack in there. I can move it up and down at least a couple of inches - should only be about half an inch at most.

M: So what needs to be done?

T: I'm afraid it'll have to be redone with new cable.

M: Can't we use the existing stuff?

T: No, there isn't enough slack to make a loop at the end.

M: OK, if you say so. Is there anything else?

T: Well, just a couple of points about this new plastic conduit over here.

M: Oh yes, I had the old one replaced when the radio was put in.

T: Right, well, it's OK except for two things. First, these drainage holes at the bottom are a bit roughly finished - there are quite a few burrs on them. And the other thing is the position. I think it could be moved further back. At the moment, it's sticking out too much. I can see your passenger grabbing hold of it by mistake.

M: So is that difficult to fix?

T: No. I'd say it can be removed, the edges of the hole smoothed, and then repositioned in a safer spot, no problem.

M: Oh well, that's something, I suppose. How much do you think it's likely to cost?

Unit 9, Lesson 5, Track 36

B: Hello, this is Bill Williams and welcome to the 12 o'clock news spot on radio 310, where local news comes first. Well it was a very unlucky day for several people out at Wageroo airfield this morning when Mike Grigson lost control of his plane on the ground and it went charging off all over the tarmac, finally ending up in the hangar, where the air taxi just happened to be in the way.

And we've put a picture of it on the website if you can bear to see the results. How did he lose control you want to know. Well, it seems Mike wasn't actually in the plane at the time - you couldn't make it up if you tried. Mike's on the line right now in fact. G'day, Mike?

M: Yeah, hello Bill.

B: How are you feeling about all this?

M: Er, most of all, I'm just so relieved that no-one was hurt. And of course I feel like a complete idiot. It just goes to show you're never too old to make silly mistakes.

B: Tell us what exactly happened.

M: Well, I was taking my young brother up for a ride this morning - lovely flying weather it was. We've got a two-seater single-engined thing. Light aircraft -

B: - it's a beautiful day yeah -

M: - yeah and ... but, the engine wouldn't start on the electric starter. I'd run a couple of electrical checks and discovered it was the battery. I hadn't been up for about six weeks, so the battery had self-discharged – it was flat, there wasn't enough juice in it to turn the engine over. It's been really hot the last couple of weeks and that makes them run down faster ... my own fault really. I should have taken the battery out yesterday and charged it up overnight. There's a good charging set in the hanger. Anyway I hadn't, so -

B: Right, so you decided to try a bit of hand propping then?

M: Yeah, I mean it's not ideal, but without an EPU, an electric, an external power source – um, unit – and once you get the engine started, the electric starter motor turns into a generator and starts charging the battery up very quickly. So when you start the engine next time – no problem. Yeah, so I left my brother in the cockpit in charge of the brakes and gave the propeller a pull.

B: He's a qualified pilot, too, is he?

M: Well, no, he's had a few lessons, but he hasn't got a licence.

B: OK, so it was always going to be a risky procedure then.

M: Yeah, well it is, you just grab the prop and swing it round by hand to turn the engine over.

B: You got the engine started though - obviously?

M: Yeah, first time. I think that was the problem actually. Tom wasn't expecting the engine to start so quickly. He got a bit scared and forgot what I'd told him and basically lost control, and I just jumped clear and he taxiied off into the hangar. Next thing I heard was this terrible grinding noise.

B: Which was your plane running into and chewing up the other one?

M: Yeah. Then the engine cut out and there was this awful silence.

B: And nobody hurt - that's amazing really.

M: Yeah only my bank account. It's going to cost me a lot to pay for all the damage - and my pride.

B: It does look a bit like a sliced loaf in the photo. Nice plane, too.

M: Don't remind me. Twin engined four seater. Rear fuselage and starboard wing ...

B: So have you got any advice for any light aircraft pilots who might be listening?

M: Yes, two things. First, cancel your flight, rather than hand propping. It's not worth the risk. And second, get an alternator fitted to your plane if you haven't got one.

B: Why's that?

M: They're much better at charging the battery up fully than standard generators like I had.

B: Oh?

M: Well yes, without getting too technical, a generator can charge up the battery, but alternators are better because they charge it up even when the engine's just ticking over. You're less likely to have starting problems with an alternator.

B: OK. Well, I'm no mechanic but that sounds like good advice. Mike - best of luck with all that, thanks for talking to us. Mike Grigson - not a lucky man. On now to a story from the other end of the district ...

Unit 9, Lesson 5, Track 37

i) The engine didn't start on the electric starter.

ii) It's been really hot the last couple of weeks and that makes them run down faster.

iii) The electric starter motor turns into a generator and starts charging the battery up.

iv) Get an alternator fitted to your plane if you haven't got one.

v) I hadn't been up for weeks so the battery had self-discharged – it was flat.

vi) We've got a two-seater single-engined thing. Light aircraft

Unit 9, Lesson 7, Track 38

A: This figure shows the actual layout of the copper conducting tracks on a printed circuit board, with the holes for component leads to be inserted, viewed from the underside of the board. It is used in the production of the printed circuit board.

B: This figure shows the actual components viewed from the top of the board, as well as indicating the track routing on the underside of the board, as if the PCB was transparent. It would be used by service technicians in identifying the location of components.

C: This figure is a 3-dimensional picture of an actual circuit board, showing the components used for this circuit.

D: This figure shows the electrical connections and component values in a way that makes it clear how the electronic system operates. It is used by technicians troubleshooting the board.

E: This figure shows the system as a whole broken down into its main functional sub-systems or blocks. This indicates the way that the sub-systems relate to the whole and gives a simplified idea of the operation of the system.

Unit 9, Lesson 8, Track 39

A: Jacksons ...

B: Oh hello ... I'm just phoning to see if I can order some things from your catalogue, but I need a bit of advice if you can, you wouldn't mind explaining a few things.

A: We can't take orders on the phone I'm afraid, but I'll certainly try and give you any help you need.

B: Oh, I see - so how do I go about ordering then?

A: Online. Or by post. Do you need a copy of the catalogue?

B: No, I've got the website up on the computer here.

A: Oh good, that's fine. Well, if you click on 'products'.

B: Yep ... OK, done that.

A: Now, you've got two options - 'stock list' and 'images'.

B: Right, so 'stock list'?

A: No, 'images' is a better bet if you're not sure exactly what you want.

B: OK right, I've got a picture of each item with a number next to it.

A: Yea that's it. And to order, you just double click on the picture and it goes in a basket. You fill in your details at the end.

B: OK, well that's easy enough.

A: So, how can I help?

B: Well, I've got to do some electrical work on a small second hand plane. I've got some general tools, but I'm going to need a few more to do the job properly.

A: OK, what kind of current and voltage are we talking about?

B: It's 28 volts DC for the heavy work, plus an AC system for the avionics.

A: What voltage?

B: 115.

A: OK, well for a start, I'd recommend a voltage tester. 115 volts can give you a pretty nasty shock. Before you touch anything, you can just whip it out of your top pocket and check. They're dirt cheap and it could save your life. We do them for 115 volts, as well as 220 volt mains.

B: OK, so double click on that - the red one, yes?

A: That's it ... and while you're at it, I'd recommend you get a cheap analogue multimeter, too, if you don't already have one. They aren't quite as accurate as the good digital ones, but they're very simple, and you can always buy another one if you lose it. They're so cheap. We do the small blue one there. It comes with a separate set of leads.

B: Good idea, yeah, I can borrow a decent digital one it if I need to. OK and I need a good quality tape measure - mine hasn't been the same since I stepped on it.

A: I suppose not. Well, we do a wide range, but I'd go for the combi - you can read off the tape directly or use the digital reading at the side.

B: Oh, yes I see it. Good, yes. One of those.

A: They are very damage resistant, too.

B: OK.

A: And remember the old saying.

B: What's that?

A: Measure twice -

B: - cut once, yes. Absolutely ... Right, now let's see. Yes, soldering.

A: Are you going to be doing much?

B: Well, as far as I can tell from the manual, it's just a few joints. Most of the connections are mechanical.

A: Nuts and washers, crimped terminals ...

B: Yes, that's it.

A: OK, well you're probably OK with the simple standard one then, the small one with the red handle there, and you really need the safety stand just to avoid accidents. You'd be amazed how many happen like that. Plus a reel of solder wire.

B: Right, there we go. Erm ... actually, I don't know. I think I might be better off with this solder gun next to it. Is that a set of different sized bits with it?

A: Yes – it does give you that flexibility - probably a bit safer, too, for working inside a plane.

B: Right, so do I just unclick those two - ? – OK, right – and... click on that one.

A: For the terminal nuts, if you've got a lot of them to tighten I'd consider a socket set. It'll save you hours of work and you get a good close contact.

B: I was going to try using the ones I've got for the car. What do you think?

A: No, too big. I'd recommend the small set we do. That'll cover all the sizes you need.

B: Right, OK, got that. Now, I've got an ordinary

pair of pliers, but ...

A: Are you going to be working alone?

B: Well, yes, most of the time. Why?

A: Then, I'd really urge you to get some locking Multi-Grip Pliers. It's like having a second pair of hands. You can grip something and then free both hands to work on it. They're brilliant - I use them all the time.

B: Are they the thing that looks like a tin opener with black jaws down at the bottom?

A: That's it. And I'd recommend the single side cutters, and the long nose pliers. They're both vital really, and the single ones are better quality than the ones that come in sets ...

B: OK, so that's these three 1 ... 2 ... and 3. OK, next on my list is a really good crimping tool, I'm going to be doing a lot of that.

A: Do you see the one bottom left with the orange handles? That's your best bet. It's pretty much the one everybody uses. People swear by it – lasts for years. And if you need wire strippers, I'd go for the automatic ones - they're just next to the crimper.

B: So those ... and those ... Now what about cable and conduit work? Some of it's quite specialized.

Unit 10

Unit 10, Lesson 2, Track 40

A: OK. So you've been through the controls and you seem happy with that. We can run through them again if you like.

B: No ... no ... I think I've got it, thanks.

A: Good, good ... OK. So now before we actually go up in the air, I need to go through the flying instruments with you.

B: Right. There aren't very many are there? I thought there would be more.

A: Well, you can have more, of course, a big airliner has hundreds - but these are what we call the basic 6, plus this magnetic compass over here. In a small plane like this you can fly perfectly well with just these.

B: Right.

A: Now the first thing to say is that what you see, the layout of the panel's likely to be the same in any small plane you fly.

B: So it's easier to adapt if you change to flying a different plane?

A: Exactly. And then the second thing to realize is that part of the skill in being a pilot is remembering to check all your instruments continuously. Just the same as in a car you keep looking at the mirrors, the speedometer then the road and other traffic and back to the mirrors and so on. You can't look at an instrument and then switch off.

B: OK. So that's another good reason for having the same configuration in different planes.

A: Yep. It makes it less likely you'll mistake one instrument for another. Anyway, let's start on the top right here. That's the altimeter. And what this is, basically, is a barometer marked in thousands of feet rather than inches of mercury. It measures the air pressure and that tells you how high above sea level you are flying.

B: How does it work?

A: Well, it just weighs the amount of air above the plane. Obviously the higher you go, the less air there will be, so the weight of air - the air pressure - will be less.

B: Well that seems pretty clear.

A: Hmm. The key thing to remember is that the altimeter has to be set before you take off, local sea level air pressure can change from day to day so all the pilots in that area need to have their instruments set to the same correct sea level pressure.

B: So if I fly into a different airspace, I might need to re-calibrate it.

A: Yep, you've got it. Air traffic control will keep you informed on that.

OK, now this one in the middle, this is the attitude indicator … and that line is the artificial horizon. If you're in clouds or for some other reason you can't see out of the cockpit, well, this can be a life saver.

B: So the centre line here is the horizon?

A: Yep, that's it. And when that white bar is below it, that means your nose is pointing down, and when it turns clockwise or anticlockwise, it shows the position of the wings relative to the horizon.

B: And this one on the left is the airspeed indicator?

A: Yep. It measures the pressure of the moving air pushing against the front of the plan, and that tells you how fast you're going through the air, which is important because if you go too slow you'll stall, the plane won't stay up, and if you go too fast you'll damage the plane - you'll shake it, and break it.

B: Wooah … so it's pretty important then.

A: Certainly is. OK, now … this one here at the bottom on the left -

B: You mean under the air speed indicator?

A: Yes, that's right. It's the turn coordinator. You can use this as a back up if the attitude indicator fails. But it's really for checking that you're turning the plane properly so that you don't slip down sideways when you bank to the left or right.

B: Sorry, I don't quite get that.

A: Don't worry, I'll give you more explanation when we're airborne if it's still not clear. It'll make more sense then.

B: And this one in the middle looks a bit like a compass.

A: Yeah. It's the direction indicator, it's a gyro compass in fact - uses an internal gyroscope instead of a magnet. Magnetic compasses aren't very accurate when you're changing direction and speed or you have rough weather. When you're back on a straight level flight, you check it against the magnetic compass and re-calibrate it with this knob here.

B: OK, I see.

A: And then finally this last one on the right is the vertical speed indicator. This measures the aircraft's rate of climb or descent. In other words, it tells how fast you are going up or down. It measures the pressure change.

B: - like the altimeter?

A: Yes, that's the idea. Now have you got any

questions before we take off?

B: No I'm pretty clear, I think.

Unit 10, Lesson 6, Track 41

i)   The layout of the panel is likely to be the same in any small plane you fly.

ii)   You can't look at an instrument and then switch off.

iii)   How does it work?

iv)   You'll shake it and break it.

v)   This measures the aircraft's rate of climb or descent.

Unit 10, Lesson 6, Track 42

P: Right now you've each got a circuit board in front of you and each one has a different fault. I want you to trace the faults. It's the same circuit for each board by the way, but as you can see …

C: Yeah, mine looks a real mess. I'm not surprised there's a fault. Whoever made this up had his eyes shut, I think.

P: You're right, it's all over the place. And that's one of the most important things to remember when you're making up a board, or repairing one - keep everything neat and tidy, components either parallel or at right angles to each other - much easier to work with. Now there are basically two kinds of fault that we find on PCBs. Any suggestions? Rasheed?

R: One of the component, or maybe more, maybe don't work properly and, the other problem, sometimes the board not working right.

P: How do you mean?

R: Well copper path can gets broken.

P: Yes, that's right. If the board's been mishandled or is put under a lot of stress or vibration, then it can crack. You can get the same problem if a component hasn't been soldered onto the board properly - the component may be fine but the electrical connection isn't secure.

C: I bet that's the problem with my board.

P: You might well be right Carlos. We'll see in a minute. But one of the problems with insecure connections like this is that the fault is often intermittent.

C&R: Intermittent?

P: Yes, that means it comes and goes, so sometimes the equipment works fine, sometimes not.

R: Well how you can you find the fault then, if the board is working fine?

P: The first thing is a visual inspection. Look at the board and components under a good strong light. Use a magnifying glass as well. If there aren't any obvious cracks or bad connections, the best way is to connect up the board to a multimeter and then press down with a pencil on the component side of the board in different places. If the board itself seems OK, then put some very gentle pressure on each component. That usually works.

C&R: Right/OK.

P: Now I'll leave you to it for 10 or 15 minutes and see what you come up with. Here's the circuit diagram and a list of the readings you should get at different points on the board. OK. Any questions?

R: No, that's OK.

P: Right. I'll see you a bit later.

P: So, how did you get on? Rasheed.

R: Yeah, this resistor here is the problem. I should get a reading of 25K ohms across it, but there's nothing. It isn't loose, I've checked, but there's no current. So it's definitely faulty.

P: Well done. The problem is with a component. So that's got to be replaced then. And what about you, Carlos?

C: Well, no surprise really. Is cracked copper track. I think cos the board is so badly made. You see, this resistor has been pushed in too far and the copper connection has broken away.

P: Good. Well spotted. So what are you going to do?

R: Throw it in the rubbish - it's a piece of junk. It's difficult to find faults in a badly arranged board.

C: Yeah! Good idea.

P: Yes, normally I'd agree with you, that is the best solution. Sometimes though, we might not have a replacement available. We would need to get it working again then.

C: Well, then, I remove the resistor, repair the track, I test the resistor again, then it's OK again, I think.

P: OK, that would be fine. But this time, we'll just replace it. They're not expensive, and you save a lot of time. Time's money, remember. And make sure you fit the new one neatly.

C: OK. I can do better than this one.

P: Good, I'm glad to hear it. Right, now there are some important rules you both need to know about for doing this kind of work. I'll go through them with you before you start. First of all …

Unit 10, Lesson 6, Track 43

a)   If the board is put under a lot of stress, it might crack.

b)   One of the problems with insecure connections is that the fault is often intermittent.

c)   Throw it in the rubbish.

d)   Press down with a pencil in different places.

e)   If the board itself seems OK, put very gentle pressure on each component.

f)   Here is the circuit diagram and a list of the readings.

Unit 10, Lesson 8, Track 44

L: …, which brings us to radio navigation aids. Now in fact, radio navigation aids were developed around the same time as the mechanical aids we've been talking about - in 1926, successful two-way radio air-to-ground communication began, and the first transmitter/receiver went into mass production in 1928. The earliest radio navigation aid was called the four-course radio range, which was first used in 1929. How the system worked was this - you can see here the layout and how it works. Four towers were set in a square like this, each transmitting the letter A or N in Morse code. A pilot flying toward the square along one of those four paths would hear only A or N in Morse code, in dashes and dots. The dashes and dots got louder or more faint as he flew, depending whether he was flying toward or away from one of the transmitters, which gave him, effectively, a kind of distance reading. Then, if he turned right or left, he would soon start to hear the other letter being transmitted, telling him which quadrant he had now entered. So he also had a reading for his direction.

The first radio-equipped airport control tower was built in Cleveland, Ohio, in 1930, with a range of 15 miles (24 kilometres). By 1935, about 20 more towers had been built. Now, based on continuous radio reports from the pilot, a controller, an air traffic controller, could follow each plane with written notes on a position map. The controller would be able to clear an aircraft for takeoff or landing, although the pilot still decided on the best flight path for himself.

Now with regard to the types of radio used, until World War II, which didn't start until 1939 of course, radio navigation relied on low frequencies similar to those of an AM radio. After the war, higher frequency transmitters, called the Very high Frequency Omni-directional Radio range or VOR, further refined the early 'four-tower' concept of allowing pilots to fly inbound or outbound along a certain quadrant on a line called a radial. These VOR transmitter locations are all printed, with their frequencies and identification Morse codes, on navigation charts.

Before World War II, although as I say, pilots were in constant radio contact with a control tower, the Civil Aeronautics Administration also relied on pilots to radio their position relative to known navigation landmarks to keep aircraft safely at a safe distance from each other in flight. During the war, Radio Detection And Ranging (RADAR) was tested. Radar's primary intent was, and still is, to keep airplanes safely separated – as you know, radar uses the delay between sending a radio ping and receiving the bounced echo to calculate the distance to the other object. And it worked, still works, very well, indeed for that. But it was not designed to guide aircraft to a specific point. Because one important thing about these various radio-based systems is that they are sufficient for navigating between airports, but they are called non-precision aids because they are not accurate enough, and don't provide enough information, to allow a pilot to land.

So the question then is how do pilots get enough information to land safely? Well, of course the principal considerations for landing an aircraft are …

Unit 10, Lesson 10, Track 45

G: Are you still working for the engineering company?

G: No no, I moved in the end to an aircraft maintenance firm at the airport. It's smaller and friendlier than the last place, and I'm in charge of all the -

D: Dad! Dad, look what I got.

H: That looks interesting.

D: It's an old coin. I think it might be really old - Roman or Greek or something.

H: David's trying out his new metal detector. David, this is an old friend of mine, George Martin.

G: Hello, David. Pleased to meet you.

D: Hello.

G: Is that a new machine?

D: Yes. I got it for my birthday.

G: What frequency does it operate on?

D: Uuuhmm … I don't really know how it works.

G: You know what an alternating current is?

D: Yes, electricity that goes backwards and forwards in opposite directions. We did that in physics at school.

G: OK, well, that's exactly what your metal detector does. It's got an oscillator circuit that makes a high frequency AC signal that goes to the coil at the end of the rod, the search coil.

D: Is that what I can hear in the headphones - the oscillator?

G: Yes, that's the noise you hear. And when you pass that coil over a piece of metal, the electromagnetic field in the coil induces a current in the metal. It's called an eddy current. And that

makes another magnetic field, which your detector picks up. Your machine just listens for electromagnetic fields.

H: And it changes the sound he hears in the headphones?

G: Yes, that's it, and that tells you there's something metallic nearby. We use something very similar at work. We've got instruments that work in the same way.

H: Metal detectors?

G: Not exactly, although metal detectors like this are what they use for security at airports, of course. No, but we use various instruments to check for faults - cracks or corrosion or even invisible voids inside the metal parts of the planes. Particularly bits that get a lot of wear and tear - failure would be very dangerous.

H: Like the landing gear -

G: - or the engine mountings, yes.

D: Why can't you just look for them?

G: Well, because quite often they're hairline cracks, very tiny and often hidden under paint or dirt.

H: So how do you detect the crack?

G: Well, if there is a crack or some other kind of damage to the structure of the metal, the current won't flow through it as easily as it does through a normal piece of the same metal.

D: So, you get a strange sound through the headphones?

G: Well our instruments – they're called eddy current meters, have a display rather than headphones, and the instrument we use out on the airfield is quite small, about the size of a very big mobile phone, but yes, you get a different reading. You use a ref- what's called a reference sample, a piece of metal you know is sound, and take a reading from it, and then see if there's a difference between the reading from the part you're testing and the reference reading. Do you see what I mean? If there's a difference it means there's some kind of fault in the metal.

H: So someone can go and fix it.

G: Or replace it, usually. It's quicker.

H: Right … so metal detectors aren't just …

D: Dad … I'm just going to do a bit more before it

starts raining again. There might be some more coins over there.

H: Righto, see you later. We'll take that coin to the museum this afternoon.

G: Good luck, David!

D: Thanks. See you later, Dad.

Unit 11

Unit 11, Lesson 4, Track 46

M: Badly designed maintenance documents can be a cause of mistakes. Written procedures that can have more than one meaning, or are long, wordy and repetitive are likely to cause confusion. Nowadays, many aviation documents are being written in a special kind of English language called Simplified English to make texts as short, simple and clear as possible. This makes the language of maintenance documentation more accessible, particularly for personnel who use English as second language. Even small improvements in page layout, diagrams and warnings can help to reduce errors. For example, many companies print maintenance documentation in upper case (capital letters), even though it has been known for many years that such text is more difficult to read than text written in the usual mixture of lower and upper case. Replacing blocks of upper case text with normal mixed-case text can increase reading speed by more than 10%.

Unit 11, Lesson 6, Track 47

Report A:

We took off without any trouble, until we tried to bring up the landing gear. The left hand gear wouldn't come up - you know why? Maintenance had forgotten to remove the pin before they moved the plane from the hangar to the gate. Anyway, we had to divert to another airport nearby, because of heavy traffic. We landed again ok and the pin was removed. Now maintenance checked it and didn't see it – the pilot did a pre-flight and missed it, and even the push crew who moved them away from the gate failed to spot it. Now ... my guess is that there was no warning flag on the pin. I'm sure someone would have noticed if there had been. So that was two mistakes -or three if you count the inspection failures as well.

Report B:

We were flying from our maintenance base to the airport to take the plane back into service. As part of the cockpit check, I called for a fire warning system check and the Captain said "complete". Everything seemed fine. It was an uneventful flight and we handed over to a relief flight crew at our destination. But they discovered that the fire extinguisher indicator lights weren't illuminated - and in fact it turned out that the bottles weren't actually connected! They were still unplugged from when the plane was in maintenance. They'd forgotten to hook up the bottles again! I can't believe that the Captain missed it first time round ... but I suppose that's what must have happened.

Report C:

As we pushed back from the terminal at Dallas airport we got a report that we'd hit another aircraft with our tail. Maintenance was called and they gave us the all clear. Anyway, we flew on down to Mexico City, arriving there at just after midnight local time in the pouring rain. I did the pre-flight with a torch, just before we took off at 2 am – still raining hard – all looked ok. We arrived in Chicago five hours later and when the aircraft was being inspected in the hangar they found tail damage. Now I'm in trouble with management because they say I didn't do a proper inspection and I shouldn't have taken off from Mexico. It's crazy. Maintenance have all the equipment in the world and they didn't spot it first time round. I'm supposed to see it at 2 am in the pouring rain, in the dark, with a flashlight. It's ridiculous.

Report D:

I accepted the aircraft for a test flight after it had had a couple of weeks in maintenance for fuel tank leaks. The mechanic in charge told me that they'd put in 1,200 lbs of fuel per wing. The pre-flight was fine and take off normal and the short flight went OK, although I noticed the fuel transfer light came on just before landing and the fuel gauges indicated zero intermittently, but I suspected a faulty connection problem. I took it up again for a longer run and about 3 minutes into the flight, the right engine flamed out, followed 90 seconds later by the left. The engines wouldn't re-start and I was forced to make an off runway landing on a bit of empty highway. Luckily, it isn't used very much. Of course, it turned out that both main tanks were empty. I've learnt my lesson. In the future, I will watch the fuelling being done, look in the tanks myself, or insist that I have an authorised fuel delivery document.

Unit 11, Lesson 8, Track 48

The efficient maintenance of aircraft depends on the right kind of tools, equipment and facilities. Heavy maintenance, which requires the plane being taken out of service, is usually done inside a hangar. A well-designed maintenance hangar should have the facilities you can see here on the screen, some of them perhaps obvious, I think:

- heating, ventilation, and air conditioning systems
- lighting, including emergency lighting
- main, sub-main, and small power
- fire detection and alarm systems
- fire protection systems
- domestic and process water services
- process ventilation
- compressed air
- lightning protection and main earthing
- energy management

The efficient maintenance of aircraft depends on the right kind of tools, equipment and facilities. Heavy maintenance which requires the plane being taken out of service is usually done inside a hangar. However, the overall efficiency of a major maintenance hangar – whether it provides aircraft overhaul, heavy maintenance, or aircraft paint spraying – is dependent upon the specialist access and lifting equipment provided within the hangar. After the aircraft has been positioned in the hangar, the access docking must quickly wrap around, so as to enable maintenance to commence immediately.

The provision of undercarriage lifting platforms allows removal and testing of the undercarriage equipment without jacking of the aircraft. It is also used to level the aircraft with the access docking before maintenance operations can begin. After the operation is complete, the aircraft is supported by jacks along the fuselage and steadied by the wings. The platforms can then be lowered to undertake maintenance of the undercarriage. Works to all parts of the aircraft can continue to be carried out uninterrupted at a level that is unaltered throughout.

OK, so moving on to some specific pieces of equipment …

Unit 11, Lesson 10, Track 49

'supervisor

'damaged

'ordered

signed 'off

co'nnectors

e'lectrics

'structural

'generator

'battery

re'placed

Unit 11, Lesson 10, Track 50

1. So I've ordered a new pump.
2. I just replaced it. Put a new one in.
3. One of the O-ring seals was damaged and there was a leak.
4. It wasn't charging the battery properly.
5. No corrosion, no sign of any structural weakness.

6. I refitted it with new connectors.

7. The boss has signed that off, too, so that's OK.

8. Oh yes, problem with the starter generator.

Unit 11, Lesson 10, Track 51

M: Hello, Fieldings, maintenance. Mike Armstrong speaking.

D: Hi, David Greenhill here. I'm phoning to see how things are going on my plane.

M: Oh yeah. Which one was it? We've got a few in at the moment.

D: A Skybird 406 - single engine four-seater.

M: Ah, yes, the red one.

D: That's it.

M: Yep, I did a bit of work on the electrics and instruments ... just hang on a minute. I'll get the job file out. Right, OK, let's have a look. Oh yes, problem with the starter generator.

D: That's right ... it wasn't charging the battery properly.

M: Yes, actually it turned out it was just a loose connection. Anyway, I took it right out, cleaned it and refitted it with new connectors. All charging up nicely now.

D: What about the magnetic compass? I think it was broken.

M: Yeah, it was. I just replaced it, put a new one in.

D: OK, good.

M: Yes, John Maddox has okayed the avionics and the electrics. Airframe inspection's also been done. No problems there - no corrosion, and no sign of any structural weakness ... the boss has signed that off, too, so that's OK.

D: That's good news. John Maddox is your supervisor, is he?

M: Yeah ... Saved yourself quite a bit of money there ...

D: Yes. Did you do the airframe inspection?

M: No, it was Bob, Bob Higgins. It's down here he's also done the nosewheel.

D: Yes, there's been a bit of vibration from that the last few times I've been up. I should have put it in for the service earlier really.

M: Yes, the problem was the damper hydraulic fluid. One of the O-ring seals was damaged and there was a leak. Not much, just a weep really ...

D: Enough to make the landing feel pretty strange, I can tell you.

M: I bet, yeah. That hasn't been okayed yet, though. It'll have to do flight tests first.

D: Sure. And what about the engine and fuel system?

M: Right, let's see. Lutfi Tarhoni did that, but ... there's something not quite - hang on, he's just come into the office. I'll hand you over, he can give you a better idea. Lutfi, can you have a word with this gentleman?

L: Who is it?

M: Mr Greenhill - the Skybird 406. 100-hour.

L: Oh yes ... Hello, Mr Greenhill?

D: Hello.

L: Lutfi Tarhoni with you ... Yes, sir, no problem with the engine itself. I did the standard overhaul on that as per the manual ... but the fuel pump, well you can say it's still working, but only 60 or 70%, and to be honest with you, not worth repairing. So I've ordered a new pump. Which should be here Tuesday next, I'm afraid.

D: OK, well, that's alright, I'm not in a hurry. I'll call back next week then, say Thursday?

L: Yes, sure I think that will be fine. If we finish it before then, we'll call you.

D: OK, thanks very much. Talk to you next week. Cheers.

L: Goodbye.

Unit 12

Unit 12, Lesson 2, Track 52

Just to see how effective these agents are, we compared Halon with CO_2 and a dry chemical bottle. Um, now this was purely backyard pyromania - we claim no scientific basis for our tests - but the results were impressive. For a test bed, we glued upholstery fabric to foam backing somewhat similar to the material used for aircraft interior panels. We dabbed on a small glass full of gasoline and ignited the panel.

Our Halon extinguisher put out this blazing mess in

minimum time – with just a couple of squirts. There was no re-ignition uhmm after only about half of the two-and-a-half pound bottle was used. And we were happy that the extinguisher had enough for another go if needed.

Then using a fresh, identical panel, we next tried the CO_2. It took about twice as long as Halon to put out the blaze – about three or four squirts. But much worse than that, once the fire appeared to be out, it flamed right back again and needed to be smothered again with the CO_2, so that's maybe 8 or 10 seconds overall, which can be a long time in an emergency of course. By the end of the second try, the bottle was nearly empty - and this fire was not particularly large.

The several types of dry chemicals we tried were nearly as effective as Halon in extinguishing the fire – they all did the job after a couple of squirts, using maybe two thirds of the content. But the mess, the resulting mess was a sight to behold, there was powder swirling in the breeze and coating everything in sight. The air was full of a biting, sour-tasting, white dust. We could only imagine trying to fly in a closed cabin with this stuff in the air; parachutes would be preferable.

Unit 12, Lesson 3, Track 53

O: But we will make our first stop in about 15 minutes at two old houses. Now, is everyone comfortable, not too hot not too cold?

All: yeah/lovely/fine

O: Good. That's because this coach has modern air conditioning. The temperature outside today is quite high - more than 30 degrees for sure - and because we are not far from the sea, there is high humidity - but we are all comfortable … yes … because the AC takes air from outside, mixes it with recycled air from on board, cools and dries it. And what should we do if the AC breaks down?

S1: Open the windows, of course.

O: For sure, because then we get air flowing through fast because the bus is moving and the air cools us by a process of evaporation from our skin, a bit like an electric fan. But if the bus

stops we are in trouble - we'll soon get hot and humid again.

S2: So how did people stand it in the days before electricity and petrol engines?

O: Well exactly. Any ideas anyone …?

Unit 12, Lesson 3, Track 54

O: OK – are we all here? Right, now you'll see that several of the suggestions you made actually were quite right. Now as you see this first house has the main building here, the garden and then at the end of the garden, a tall tower. That is called a wind tower, and it's actually an air intake. It's built as tall as possible because the higher you go the greater the pressure, the force, of the wind, and the lower the air temperature. And the relatively cool fast-moving wind is ducted down the inside of that tower and under the ground through a tunnel which, compared to the outside air, is quite damp. As the wind passes the tunnel walls you get evaporation, the air takes heat from the tunnel, so it cools – the tunnel cools in the same way that your skin cools. This means the air coming behind will be cooled and flows into the house and out of the open windows, creating a comfortable environment. And the cycle repeats: the wind flow cools the damp tunnel through evaporation, and the tunnel then takes heat from the air following behind. You don't look convinced. Come inside and you'll see.

O: There - what did I say? Much cooler, isn't it?

S3: That's amazing! How much cooler is it in here? 10 degrees difference?

O: Yes, something like that, somewhere between 5 and 10 degrees. Low-tech, but very effective. Now we're going to go up to the top of the house and on to the roof so we can see another AC system in operation in the buildings around this one. Be careful of the stairs - they are quite narrow.

Unit 12, Lesson 3, Track 55

We'll just wait for the last few … OK. If you

look over here you can see that building opposite with the high round roof, and at the top, there is a small dome shape with holes around it. Now what happens in this case is that the wind flows over the top of the dome and as it does so it creates low pressure, like the airflow over the top of an aeroplane wing. And because of the lower pressure over the roof, the warm air inside the house is pulled upwards and out of those holes. New air flows into the bottom of the house, and you get a constant airflow through the interior.

S4: But they didn't try to cool the air like in this house?

O: Well, that house also has a big tank, a reservoir, of water under part of the main floor, to help keep the ambient temperature low. Like with the water in the tunnel in the case of this house, evaporation keeps the air temperature down. Unfortunately, they are doing maintenance work at the moment, so we can't go inside. But go and have a look at the wind tower. It's a great lesson in simplicity – the best ideas are often based on simple concepts. Do your sketches and make some notes, and in half an hour we'll go back to the coach and yes we should have time to go to the souk before we go back to …

Unit 12, Lesson 3, Track 56

a) environment / average / conditioning / contaminant

b) system / pressure / falls / intake

c) distribution / reservoir / temperature / ventilate

d) rise / ducting / moist / cloud

e) volume / cool / mixture / moisture

f) exhausted / contracting / compression / constant

g) humidity / saturate / evaporate / equivalent

Unit 12, Lesson 6, Track 57

… Can you see that OK? Now the three aircraft here, the Sunseeker, the Hughes 300 helicopter and the BAe Hawk, are unusual because they are powered by electric motors rather than by combustion engines. Of course, only this one is big enough to carry a pilot. These other two are working models, but electric motors have several advantages. They are not as noisy as other engines … they produce no dangerous sparks and, of course, they are cleaner. Not only that, they have a longer life and are far more reliable than combustion engines. And – as well as all that- they are extremely efficient and can convert much more electrical energy into mechanical energy than conventional engines.

Unit 12, Lesson 6, Track 58

Now the fact that it is now possible to fly these aircraft with electric motors is partly due to the development of the brushless DC motor. Look at this. This is a cross section. In standard brushed motors, as we saw before, the electromagnetic armature on the outside here rotates past static magnets as it is supplied with power through the brushes. But in this BLDC motor, brushless DC motor is like a brushed motor turned inside out. The electromagnets, in orange here, remain static, and the magnets – in pink - rotate on this central shaft, the black shaft. There is no direct physical connection between the power supply and the rotor around the shaft, the rotor is the grey section here. Instead, the brush and commutator are replaced by an electronic controller.

Unit 12, Lesson 8, Track 59

JB: … looking at changes in the manufacturing industry and the aircraft manufacturing industry in particular. With us, is Alan Bowden, chief engineer of Apex Aircraft, who produce small passenger jets and Turbo prop trainers. Is that right, Alan?

DS: Yes, quite correct, we've just started production of our new advanced training aircraft, the D30.

JB: Yes, I'd like you to tell us more about that. But let's go, let's go straight to our first caller, Donald Smith, who's on the line from Manchester. Donald, what's your question?

DS: Good afternoon.

AB: Hello, good afternoon.

DS: You were talking about the new plane -

AB: The D30.

DS: Yeah. Well, I'm really interested in planes and I want to do engineering at college, and I heard something about it on the radio, and they said it had a glass cockpit, as if that was something special, and I was curious. I mean, I thought all cockpits were made of glass, at least the windows are, so I'm assuming perhaps this is something different …

AB: No, no, you're right. The term 'glass cockpit' doesn't refer to the windows. It refers to the flight instrument display which the pilot uses to get information as he flies the plane.

DS: So, it's a computer system …

AB: Well, yes, there's certainly a computer - sometimes several - involved. But the most important feature is the way that the six traditional flight instruments are replaced by something called the Primary Flight Display. The PDF for short. Instead of the pilot having to look at six different instruments at once, all the basic flight information is grouped together on the same screen, which means there is less chance of him making a mistake because of misreading his instruments at critical times, like coming in to land.

DS: OK … and the PDF is made of glass.

AB: Well, actually, well, that's where the name comes from, but in fact, it's an LCD computer screen with an image of the information on it. It's called glass because the first screens were made of glass, like an old television screen.

DS: Right, I see … and all information is on there …

AB: Yes. The centre of the display shows the earth in brown and the sky in blue, with a red cross where the centre of the aircraft is. This does the job of the Attitude Indicator - which has been present in aircraft since the earliest days of flying: the gradations above and below the horizon show degrees of pitch up or down of the nose, and two black horizontal lines represent the wings of the aircraft. They show the amount of roll - whether the aircraft is flying level or not. The Attitude Indicator is in the middle because it's still considered the most important instrument. Um, for turning, in a turning situation, the pilot also has a white arrow at the top of the screen. The arrow swings left or right and allows him to monitor the turn, like a traditional Turn Coordinator. Now around that

central screen, on either side of the display, are what are known as the 'tapes,' which take the form of vertical grey panels. On the left side you have the Airspeed tape, and on the right side the Altitude tape, which has a further tape on its right showing the rate of climb or descent. That's the Vertical Speed Indicator. The tapes scroll up or down depending on speed and altitude and in the centre of each tape there is a small box highlighting current airspeed - on the left - and current altitude on the right. And you've got a Gyrocompass at the bottom of the screen to give you your heading.

Unit 12, Lesson 8, Track 60

JB: So, it's a lot like the old basic six really?

AB: Yes, only easier to use. And finally, when the aircraft is getting near the ground, when it is landing, the pilot will see a visual indication of the runway on his PFD. It shows him the position of the Instrument Landing System on the ground. This, and signals coming from the ground, allow him to land using this instrument alone - for example in bad weather when he can't see much outside the aircraft.

DS: And what else is there besides the PDF?

AB: Well, there is a lot of information about the aircraft systems of course, which is displayed electronically, but the next most important instrument I suppose is the Navigation Display or ND as we call it. It combines the functions of radio navigation instruments such as the ADF indicator with a moving map which can be made bigger or smaller.

JB: A SATNAV sort of thing?

AB: Yes, in a way, plus it can also show the shape of the land you are flying over as well as landing approach maps, weather maps and 3-dimensional vertical displays, and 3D navigation images. And the great thing is, the pilot can control a lot of the information, he can change the way it is displayed, or even remove it sometimes. The glass cockpit is actually a result of the same technology that goes into Satnavs, digital radios and home computers. It's a combination of the three things really. Global Satellite Technology for navigation information, lightweight LCD display screens and powerful small electronic computers to link everything up. It's a real improvement.

JB: Right ... Thank you for that ... I hope that answers your question Donald?

DS: Yes, thanks very much ... I think I got the general idea.

JB: Good. Now, Alan you've been doing a lot of work on the new D30 design. Tell us ...

Unit 12, Lesson 10, Track 61

T: OK, there aren't any obstructions around the aircraft.

E: Right. What's next?

T: Number two: 'Position the tail stand under the tail jacking point. Raise the stand central pillar to the approximate height of the jacking point and insert the locking pin through the centre tube and pillar. Adjust the tail stand threaded ball-end until it contacts the jacking point, ensuring that all three legs of the stand are firmly in contact with the ground' ...

E: OK, good, that's fine. Now as it says, we need a jacking adapter on top of each jack, which we've just done, and then position them under your jacking points. Make sure the hydraulic valves are closed for now. OK, closed? And the other one. Right, raise the jack ... that's it ... until the adapter just touches the jacking point ... Good. Stop there. Now, line up the adapter and your jacking point properly. Right. And do the same on the other one.

T: It says 'slowly and evenly' - is this OK?

E: That's fine. Keep going. Till you see the wheels are just off the ground.

E: Right, now it's not all that heavy but we need two pairs of hands for this. Don't try and do this on your own, the manual says. I'll lift the tail and you adjust the stand again to the right height and put a locking pin in. Then I'll put the tail down on it and we'll check if it's at the right height. OK, ready?

T: Yes.

E: Don't take too long over it, mind you - it's not that light either.

T: Right, so the two main jacks have got their locking collars on, so ... 'Raise the aircraft tail', yes, I wasn't sure about this - 'finally adjusting it to the required level by use ... of the threaded ball-end.'

E: Yes, that's important. The ball-end is - look here at the tail stand. See this ...

E: That's it, slowly does it. If you let it down too fast and there's a problem it can be hard to do anything about it. It's ... it's on the collars now. Close the hydraulic valve - let me see you do it. OK. And the other side. Remember you have to do both jacks. That locks it all up - do you see how the locking collars lock the jacks mechanically, and the valves lock them hydraulically. That level of security's mandatory. Better safe than sorry.

E: What do you think of the tail height now?

T: It looks level to me.

E: Good, yes it's OK as it is.

T: I was wondering though, where do we find the forty kilos of ballast - is there a standard ready-made weight? This is the suspension point ...?

E: We'll get to that in a moment. The first thing is that that is the ballast point, yes.

CD 1

TAKE-OFF: Technical English for Engineering

Tapescript Number	Unit / Lesson	CD Track Number
1	Unit 1, Lesson 1	1
2	Unit 1, Lesson 4	2
3	Unit 2, Lesson 2	3
4	Unit 2, Lesson 4	4
5	Unit 2, Lesson 6	5
6	Unit 2, Lesson 6	6
7	Unit 2, Lesson 8	7
8	Unit 2, Lesson 10	8
9	Unit 3, Lesson 3	9
10	Unit 3, Lesson 3	10
11	Unit 3, Lesson 6	11
12	Unit 3, Lesson 9	12
13	Unit 4, Lesson 2	13
14	Unit 4, Lesson 2	14
15	Unit 4, Lesson 5	15
16	Unit 4, Lesson 6	16
17	Unit 4, Lesson 8	17
18	Unit 4, Lesson 10	18
19	Unit 5, Lesson 2	19
20	Unit 5, Lesson 2	20
21	Unit 5, Lesson 6	21
22	Unit 5, Lesson 8	22
23	Unit 5, Lesson 10	23
24	Unit 5, Lesson 10	24

CD 2

TAKE-OFF: Technical English for Engineering

Tapescript Number	Unit / Lesson	CD Track Number
25	Unit 6, Lesson 4	1
26	Unit 6, Lesson 8	2
27	Unit 7, Lesson 2	3
28	Unit 7, Lesson 4	4
29	Unit 7, Lesson 8	5
30	Unit 8, Lesson 1	6
31	Unit 8, Lesson 3	7
32	Unit 8, Lesson 5	8
33	Unit 8, Lesson 5	9
34	Unit 8, Lesson 9	10
35	Unit 9, Lesson 3	11
36	Unit 9, Lesson 5	12
37	Unit 9, Lesson 5	13
38	Unit 9, Lesson 7	14
39	Unit 9, Lesson 8	15

CD 3

TAKE-OFF: Technical English for Engineering

Tapescript Number	Unit / Lesson	CD Track Number
40	Unit 10, Lesson 2	1
41	Unit 10, Lesson 2	2
42	Unit 10, Lesson 6	3
43	Unit 10, Lesson 6	4
44	Unit 10, Lesson 8	5
45	Unit 10, Lesson 10	6
46	Unit 11, Lesson 4	7
47	Unit 11, Lesson 6	8
48	Unit 11, Lesson 8	9
49	Unit 11, Lesson 10	10
50	Unit 11, Lesson 10	11
51	Unit 11, Lesson 10	12
52	Unit 12, Lesson 2	13
53	Unit 12, Lesson 3	14
54	Unit 12, Lesson 3	15
55	Unit 12, Lesson 3	16
56	Unit 12, Lesson 3	17
57	Unit 12, Lesson 6	18
58	Unit 12, Lesson 6	19
59	Unit 12, Lesson 8	20
60	Unit 12, Lesson 8	21
61	Unit 12, Lesson 10	22